The International Behavioural an

THE DYNAMICS OF A CHANGING TECHNOLOGY

TAVISTOCK

INDUSTRIAL RELATIONS
In 13 Volumes

I	Efficiency and Effort *W Baldamus*
II	Interdependence and Uncertainty *Edited by Charles Crichton*
III	Form and Content in Industrial Democracy *F E Emery and Einar Thorsrud*
IV	The Dynamics of a Changing Technology *Peter J Fensham and Douglas Hooper*
V	Communications in the Building Industry *Gurth Higgin and Neil Jessop*
VI	The Game of Budget Control *G H Hofstede*
VII	The Changing Culture of a Factory *Elliott Jaques*
VIII	Measurement of Responsibility *Elliott Jaques*
IX	Mid-Career Development *Robert N Rapoport*
X	The Enterprise and its Environment *A K Rice*
XI	Managers *Kenn Rogers*
XII	New Ways in Management Training *Cyril Sofer and Geoffrey Hutton*
XIII	The Enterprise in Transition *H van der Haas*

THE DYNAMICS OF A CHANGING TECHNOLOGY

A Case Study in Textile Manufacturing

PETER J FENSHAM AND
DOUGLAS HOOPER

LONDON AND NEW YORK

First published in 1964 by
Tavistock Publications (1959) Limited

Published in 2001 by
Routledge
2 Park Square, Milton Park, Abingdon, Oxfordshire OX14 4RN
711 Third Avenue, New York, NY 10017

First issued in paperback 2014

Routledge is an imprint of the Taylor and Francis Group, an informa business

British Library Cataloguing in Publication Data
A CIP catalogue record for this book
is available from the British Library

The Dynamics of a Changing Technology
ISBN 0-415-26439-1
Industrial Relations: 13 Volumes
ISBN 0-415-26510-X
The International Behavioural and Social Sciences Library
112 Volumes
ISBN 0-415-25670-4

ISBN 13: 978-1-138-86373-6 (pbk)
ISBN 13: 978-0-415-26439-6 (hbk)

The Dynamics of a Changing Technology

A CASE STUDY IN TEXTILE MANUFACTURING

PETER J. FENSHAM
DOUGLAS HOOPER

Foreword by O. L. Zangwill

TAVISTOCK PUBLICATIONS

First published in Great Britain in 1964
by Tavistock Publications (1959) Limited
2 Park Square, Milton Park, Abingdon,
Oxon, OX14 4RN
This book is set in 12 on 13 point Bembo

Contents

Foreword by Professor O. L. Zangwill *page* ix

Acknowledgements xv

1 Introduction 1

2 The Textile Company 12

3 Debenham Mill 26

4 Radbourne Mill – The Coming of the Automatic Looms 40

5 Management Under Change 67

6 Changing Tasks for Operatives 90

7 Selection and Training for Innovation 127

8 Changing Relationships 147

9 Communication 179

10 Conclusions 206

I *General* 206

II *Implications for Management* 217

III *Social Science Concepts and Theories* 224

General References 233

Index 237

Figures

1 Layout of Debenham Mill *page* 27
2 General organization of Debenham Mill 28
3 Formal management organization of Debenham Mill 31
4 Layout of Radbourne Mill in mid 1954 41
5 Formal management organization of Radbourne Mill in mid 1954 42
6 Efficiency and quality of production in the automatic section 62
7 Total production of cloth in the automatic section 62
8 Turnover of warps and loom downtime in the automatic section 63
9 Reorganized layout of Radbourne Mill by early 1956 64
10 Formal management organization of Radbourne Mill in early 1956 after the innovation 64
11 General management organization of Radbourne Mill 69
12 Production of warp and weft at Radbourne Mill 76
13 Interpersonal interaction of Radbourne Mill management 83

Tables

1 Weekly loom efficiency, Debenham Mill *page* 30
2 Interaction at daily meetings, Debenham Mill 34
3 Total number of looms per operative 59
4 Chronology of the significant changes in the automatic section during the innovation period 66
5 Production potential, Radbourne Mill 70
6 Mean downtime of looms per warp changed: July–October 1955 79
7 Work-study components of weaving 92
8 Time spent as trainee weavers 138
9 Non-automatic weaving operatives' time ratio and loom coverage 149
10 Automatic weaving operatives' time ratio and loom coverage 153
11 Debenham Mill: Non-weaving operatives' attitude to communication 190
12 Radbourne Mill: Non-weaving operatives' attitude to communication 190
13 Debenham Mill: Weaving operatives' attitude to communication 193
14 Radbourne Mill: Weaving operatives' attitude to communication 194

Foreword

The authors of this book describe in some detail the effects of an important change in production technology in a British textile company. They are concerned not with the technical or the economic aspects of this change but solely with its psychological repercussions. In a study lasting a little over two years, the authors made a determined attempt to understand what modern technological change really means to those whose day-to-day work is directly affected by it. They have tried to show that changes in technology cannot be divorced from their social context and that social attitudes must necessarily undergo change if technological innovation is to run smoothly.

The book is the outcome of a collaboration between an Australian physical chemist, Dr Peter Fensham, and a British social psychologist, Dr Douglas Hooper. Dr Fensham was awarded a studentship by the Nuffield Foundation in 1953 to undertake training and research in the Social Sciences, based on the Cambridge Psychological Laboratory. Dr Fensham's search for a bridge between the exact sciences and the inexact science of what might be called 'real' (and more complex) human issues, led to a joint exploration, on the part of the two authors, of the field of modern social psychology, particularly in its experimental aspects. Without wishing to disparage the work of those who have sought to develop the subject as a laboratory discipline, they came to the view that the best approach lay in fieldwork illuminated by some understanding of general psychological principles. Although lacking in experimental controls, such work can come to grips with 'real-life' issues and can define areas in which a more precise laboratory analysis would be fruitful. The questions therefore arose as to what types of problem would best repay intensive observational study.

Dr Fensham expressed the view that he would prefer to work in the general field of industrial social psychology, if possible on a problem connected with the increasing application to industry of automatic control procedures. Accordingly, we submitted a research project to the Committee of Human Relations in Industry which had been established shortly before under the joint auspices of the Department of Scientific and Industrial Research and the Medical Research Council. Our project was concerned with the general problem of modern technological change and its human implications. We made it clear that our project was to be envisaged as exploratory, its intention being to define problems appropriate to more detailed investigation rather than to attempt to resolve immediate practical issues. What was needed, we argued, was an accurate picture of what really happened in a factory undergoing rapid and substantial change in production technology.

After some preliminary discussion, our project was duly approved, and its subsequent development was very largely due to the assistance so generously provided by the Joint Committee. Dr Fensham started active work early in 1954, and his first enterprise was to visit a number of firms to ascertain whether major changes in production technology were impending, and if so, whether he could carry out an investigation of the transition period. The industries covered by this initial survey included oil refinery, light and heavy engineering, textiles, chemical manufacture, automobile manufacture, and laundering. This preliminary survey gave much valuable information – not least of the variety of meanings attached to the term 'automation' – but only one company appeared to offer really good prospects for a systematic research project. This was a major branch of an important textile concern. On the occasion of Dr Fensham's first visit, the managing director opened the interview in the following words: 'If you are interested in the social impact of automatic methods, you could not have come to a better industry. The possible automatic changes in this industry are so great that all textile companies are having to face such a change; but they are confronted all the time by social factors of a general character as well as by those of a more specific character arising from particular local conditions.' This wise and farseeing judgement set the stage for what was to become the major scene of our inquiry.

It was immediately plain that the managing director of this company was sympathetic in principle to the possibility of an outside investigation being carried out in two of the mills under his control. In the light of his encouragement, Dr Fensham was able to make the acquaintance of the general manager and the personnel manager and to visit the two mills. The next step was to draw up a programme of investigation, providing an outline of the object and scope of the proposed work and defining the status of the investigators as independent outside observers. This was duly submitted to the company. The response was an invitation to Dr Fensham and myself to discuss our proposals in full with the managing director and his senior colleagues. This meeting took place in June 1954, and was conducted in the most amicable terms. Much helpful advice was forthcoming from all present and, to our great satisfaction, the project and its conditions were accepted almost as they stood.

The setting up of the inquiry was a matter calling for considerable finesse. Wisely, Dr Fensham started with the managers of the two Mills and followed up this approach along the lines of the formal structure of the company's organization. Further, he approached the Works Council at the mill in which the major changes in production technology were impending, for their approval. Having received his reassurance that the project was based on a University Department and wholly sponsored by outside bodies they too offered no objection. With characteristic tact, Dr Fensham further offered to submit monthly reports to the Works Council – a procedure that was faithfully carried out throughout the period of the inquiry. Representatives of the two main unions at each Mill – the Transport and General Workers Union and the National Association of Power Loom Overlookers – were also advised about the proposed work and they gave formal approval in relation to their own members. General approval being then secured, Dr Fensham set about gaining some knowledge of the technology of weaving and its allied processes before beginning his inquiry into social factors.

In January 1955, Dr Fensham was joined by Dr Douglas Hooper, then a recent graduate in Psychology of the University of Reading. Dr Hooper brought to the work a wide and varied occupational background and special interest in the social aspects of his subject. From then on, the two investigators worked together in the closest understanding and cooperation until the active fieldwork was brought to a close in 1956. Several months were then devoted to an analysis of the material, part of which was written up by Dr Fensham as a thesis submitted to the University of Cambridge for the Ph.D. degree. An extensive report on the more technical aspects of the work was also prepared and submitted to the company.

All of us who were associated with this work have been in agreement that a book might one day be written about it. The actual preparation of this book has however been delayed by a variety of circumstances, not least by Dr Fensham's return to Australia in August 1956, and the demands of his present duties in the Department of Chemistry at the University of Melbourne. Fortunately, Dr Fensham has found time and opportunity to contribute his share of the writing and has been closely consulted at all stages as regards the organization of the book. To Dr Hooper, however, has fallen the main burden of the final preparation of the text for publication. In spite of their geographical separation, the two authors have worked throughout in so close an association that the present book may truly be viewed as their joint achievement.

Among the many important issues arising from this study, two in particular appear to me to warrant the most careful consideration. In the first place, it is often said that the industrial worker is 'resistant' to technological change. By this is apparently meant that the worker is actively hostile to changes in his established routine and work methods. Although this may well be true in some cases, the findings reported by Dr Fensham and Dr Hooper make it clear that 'resistance to change' is often more apparent than real. Their work suggests that much so-called 'resistance' arises less from positive disinclination to accept a new technology than from the unwitting persistence of attitudes and value-systems appropriate to the old. Attitudes are changed less rapidly than actions; with the best will in the world, it may take the operative weeks or even months to re-cast his long-established standards of skill, status, and responsibility in accordance with the demands of a new technology. This is not 'resistance' in any active sense; it is simply the time-lag inseparable from all human adaption at the social level. This time-lag may perhaps be reduced by fuller understanding of the psychological and social adjustments involved. But its existence should be borne most carefully in mind in a period of rapid technological development.

In the second place, the authors of this book bring out most graphically the variety of meanings – often implicit – which become attached to words in common use in industry and the very real misunderstandings which may arise in consequence. This is well illuminated by their consideration of the concept of 'work' itself. At management level, 'work' is understood largely in terms of production; to the work-study engineer, it is mainly a matter of units of physical effort; to the operative, finally, 'work' is assessed in terms of the demands which it makes upon physical activity and upon considerations of responsibility as traditionally understood. It is shown in this study that automatic looms which increase the number of looms assigned to the individual operative while at the same time decreasing the amount of manual labour required of him, pose very real problems in psychological adjustment. Thus it may appear to the operative that he is now being required to do 'more work' and 'less work' at the same time.

Clearly, then, a new conception of 'work' – and with it of 'skill' and 'responsibility' – must evolve before adaption to the new procedure can be said to be satisfactory. The operative has to learn that inspection activity is also a form of 'work', no less exacting or dignified than the manual weaving procedures upon which his former status in large part depended. This kind of learning cannot take place overnight: it will occur only as fresh attitudes towards the entire socio-technical situation gradually develop.

In this book there are few figures and practically no statistics. This omission may be thought in some quarters to indicate a lack of concern for scientific method. Such a judgement, however, I would regard as both hasty and ill-conceived. The object of scientific inquiry is, of course, to provide general explanatory hypotheses open to verification by experiment. This obviously demands that the data of observation should be presented wherever possible in quantitative form. Yet in psychology, alas, quantitative treatment has up to the present found but limited application. No really adequate methods are available for quantifying observations of human behaviour in complex social settings. Until the subject has advanced further, therefore, I cannot believe that anything is to be gained by dressing up social research in the superficial trappings of quantitative science. After all, the origin of all science is honest and devoted observation. If social psychology bases itself firmly on this foundation I do not believe that it can go far wrong.

O. L. ZANGWILL

Psychological Laboratory
University of Cambridge

Acknowledgements

It is some years since the work described in this book was finished, and many of those who helped us in matters great and small are no longer in the same offices and institutions. Nevertheless, our indebtedness remains and we feel that it is still appropriate to mention them as if they were still in their previous positions.

Our prime gratitude is to the foremen, workers, managers, and directors of the factories in which we worked. Our presence was often disturbing to them; sometimes we were glad to know that it was enlightening; but always, we were treated like people trying, like themselves, to do a job of work which relied – like theirs – on the cooperation of many others. It would be not only trite but untrue to say that publication of this book is their reward. But if this book contributes to understanding in any sphere, then the reader too owes a debt to these men and women.

We hope, of course, that our friends in the company will enjoy this account of their experiences. The whole material was discussed with workpeople at every level, and all personal and impersonal statements about the changes were agreed with the people directly concerned. Our analysis of situations was not always accepted, but wherever there was a difference of opinion we have either modified the text when we could agree to the objection, or else abided by our position when we felt that our own view was the more valid. In spite of such differences, our mutual respect remained.

We would like to acknowledge the direct help which we received in organizing the research plan and implementing the subsequent writing. Professor O. L. Zangwill welcomed us in turn to Cambridge and provided us with active support throughout, as his own preface testifies. In addition to him; Mr D. Marples and Professor M. Fortes at Cambridge, provided much helpful comment and interest. In the wider sphere, Mr J. R. Gass and Mr R. G. Stansfield, both of the Department of Scientific Industrial Research, smoothed our path in many direct and indirect ways.

At a later stage, there are those who worked directly towards the production of the manuscript. Mrs P. French and Mrs E. Mitchell were successively project secretaries who willingly produced the drafts and reports for us. Mr Eric Trist of the Tavistock Institute read the whole manuscript critically but encouragingly and the quality of the book has been greatly increased by his suggestions. Mrs Dorothy Paddon, also of the Tavistock Institute, helped the publishers to convert the typescript into its final form.

Finally, we are grateful to the Department of Scientific and Industrial Research for a grant which supported the study, and Peter J. Fensham records his deep thanks to the Nuffield Foundation for the award of a Sociological Scholarship which made his entry into the social sciences possible and so enjoyable.

PETER J. FENSHAM
DOUGLAS HOOPER

January 1964

ONE

Introduction

One of the main features of modern industrial society is the repeated incidence of technological innovation. This may consist of new machines, new methods of production, or new products. It may occur gradually on a small scale, or rapidly on a grand scale. The importance of human factors in such situations is often stressed; but scientific studies of these factors are still comparatively rare.

This book describes a case study of the innovation of automatic looms in a weaving company. In the industry this type of innovation is referred to as 'a change from non-automatic to automatic production'. The innovation was quite extensive and rapid and is reported as we observed it from July 1954 to September 1956. Our observations were made during long periods of attendance within the factories of the company. Our position in fact resembled that of a social anthropologist in his study of a primitive society.

We were formally accepted on our own request as independent research workers by the company directorate, the individual factory managers, the Works Councils, and the two unions involved.

This formal acceptance did not mean that our position and purpose as research workers were clearly understood, nor did it mean that we were accepted for ourselves. Acceptance was only won slowly and in different ways for each individual with whom we had contact. Our research purpose was a difficult concept for many people in the factories and it was probably never clearly perceived by many of them. This difficulty is, however, a common experience in our society despite the existence of an increasing number of research workers of all types, and individuals who accepted us for ourselves appeared to take our purpose on trust.

OBJECTIVITY AND ACTION RESEARCH

In carrying out our study, we do not pretend that our presence in the situation did not have some effect on its development. In fact, it was clear that, to varying extents throughout the period, we became part of the total industrial situation itself, and involved in its total dynamic life.

Such a surrender of objectivity, we believe, is inevitable if the dynamics of complex social systems are to be studied. The degree of involvement of the research worker can vary immensely. In the present case, our involvement was similar to and influenced by that of Ronken and Lawrence (1952) in their study of the introduction of a new product into an electrical manufacturing company in the U.S.A. That is, we did not actively avoid involvement and participation in the life of the company, provided that such action was within the terms of reference which had been accepted by all the levels in the company. Further, as Curle (1949) has put it in discussing a rationale for such 'action research', 'any action had to serve the whole factory community, and not merely a sectional interest'. Finally, it was an attempt, not to impose our own views on the factory community, but, by relieving various tensions, to release its own knowledge of its own structure, coupled with the power to act on it. However, our role differed considerably in degree from studies using field experiments. Typical of the latter are the classical Hawthorne experiment (Roethlisberger & Dickson, 1939) and the earlier and recent work of French, *et al.* (1948, 1960), Bavelas (1952), and Jackson (1953), in which planned and controlled changes are introduced into the factory situation at the direct instigation of the research worker. Such an experimental approach severely limits the extent of the situation to which attention can be given, and control of the variables over long periods of time is difficult. Furthermore, the experimenter cannot help but impose his own value judgements upon a situation in which they may not be wholly shared.

Positive involvement, but of a different character has been developed in studies by the Tavistock Institute of Human Relations through the role of the 'research consultant'. This is well illustrated by the long case study of the Glacier Metal Company which has been described by Jaques (1951). The research workers had a consultant-client relationship with the various groups in the industrial community. A group invited the research worker to help them with their problems. For example, the Works Council asked for help with certain of their procedural problems. This led to a consideration of the role of such consultative bodies, their power and authority, and an examination of the communication problems which existed. With the aid of the Tavistock consultants, the Works Council attempted to work through their problems by revealing and identifying some of the less obvious influences on their behaviour. Through such an analysis, new insights appeared which led to actual changes in the behaviour and organization of the factory community.

The involvement of the research worker can also be much less than in our case. For example, Homans (1953) in a company office, and Blau (1955) among government officials, selected for observation only a few variables, of which a measure could be obtained in an objective manner. Such variables were the mutual interactions, their friendship choices, and their job performance. An even more objective role is taken by research workers who carry out statistical studies of a large number of similar industrial groups. Data on selected variables for the groups are collected and correlations calculated between these variables. Katz (1951) in the U.S.A. and Argyle (1958) in Britain have made such studies; but, though they have a precision lacking in all the other types of study, they give little detail about the total situation, and the meaning of the correlations is not always clear.

A CASE STUDY – WHY?

The disadvantage of case studies is that they can never establish general laws or theories. But their strength is that they can reveal the important factors in complex social situations and generate powerful hypotheses. We believe that these functions are the most appropriate at the present stage of the development of a science of social behaviour. Only in such detailed observational studies is there any hope of appreciating the complexity of the social situations arising in industrial communities. Even so, there is much that is not observed; but some direct observation of reciprocal perceptions and the behaviour of the different groups of people in the situation can be obtained. Such direct observation of interpersonal and group relations goes far beyond the records that can be obtained from members of the situation in isolation.

SOME DEFINITIONS

In the description of the happenings in the company and in the analysis of these observations, we use a number of particular terms. Since most of these have been used by other authors and not always with the same meaning, they will now be defined for our present usage.

Individuals and groups are considered in a manner which stems from the field theory of Lewin (1952). That is, the *behaviour* of *individuals and groups* is determined by the interplay of influences from all aspects and areas of their life. Any things, events, objects, persons, or groups which are 'real' for an individual can exert an influence on him. Thus it is not possible to consider an individual or a group in a factory solely in terms of the events within the factory. Nor can influences from a worker's home and factory life be treated as completely separate, with neither sphere influencing the other. The different areas of an individual's life are closely linked and interdependent. What goes on in a worker's home is not only an important influence on his attitude and behaviour at work, but it also affects how he will perceive situations and objects at his work. However, in some situations this interdependence may be weak and then, for practical purposes, it is reasonable to consider only certain limited areas of influence. The influences from other areas are in these cases practically constant.

In this descriptive use of Lewin's field theory, an industrial company or factory is seen as a *social system* consisting of the interdependent social and technical organizations which include all the machines, materials, products, individuals, and groups in a dynamic relationship. In such an inextricably linked human and technical system, the introduction of some new technology cannot be represented as just so many new machines or products. It is to be seen as a multiplicity of changes in the behaviour and interrelations of the individuals and groups in the system.

An industrial company or factory as a social system has a particular character which is due to its pattern of *structure* and *culture*, on the one hand, and to the *personalities* of those involved in it at all levels, on the other. This pattern is a dynamic one which is constantly affected by the changes in the elements whose interaction produces it. At any particular time this pattern may present an external picture of apparent equilibrium with the interacting elements in dynamic balance. Even for the equilibrium patterns, structure, culture, and personality must be seen as descriptive elements and not explanatory ones. Changes in the pattern, described in terms of these elements, may at any time be initiated from either within or outside the system.

Within the factory, there are many different social structures, some of which are all-inclusive and others peculiar to two or three people. The commonly recognized structures are those known as the formal organization. This includes the hierarchies of production, the divisions of management, recognized labour unions, and such bodies as Works Councils. The position or jobs which constitute the structure are called *roles*, and these are interrelated in ways specified by the structure. These roles are not individuals but they are occupied by individuals. Thus an individual may occupy several roles in the structure, such as being the manager of the factory and the chairman of the Works Council. The role of the manager is not, however, the same as that of chairman of the Works Council. Roles exist apart from the individuals or groups of individuals who occupy them. This is clearly seen when one person occupying a certain role leaves the system and is replaced in that role by another. This distinction between roles and people is not always clear, especially when some roles in practice only exist while particular persons occupy them. Failure to make this distinction can lead to serious problems in human relations.

The relationships between roles or jobs bring the individuals in them into personal relation with each other. So also existing relationships between persons can be facilitated or hindered by the existence of their corresponding role relationships in the structure.

The *status* of a role or job is the traditional value or esteem attached to it by the members of the social system. Individuals and groups of individuals are likewise afforded value by the other members of the factory, and this personal value is called *prestige*, which attaches to individuals and not to roles.

An individual in the factory is thus regarded in a way which combines his personal prestige and the status of the job he occupies. The position of foreman has a certain status in the factory, but the regard for any particular foreman will also depend on his prestige in the eyes of others.

Each position or role in the structure carries a certain *responsibility*. This responsibility is the sum total of the work tasks, people, and equipment of which the person in that role has charge.

In any factory certain patterns of behaviour are accepted and practised. These patterns or 'total ways of life' are called the *culture* of the factory. Included in this culture are the methods of production, the traditional labour and management ways of behaving, the systems of punishment and reward which individuals and groups apply to each other, the methods of payment, and the methods of distributing work. It also includes many less formally expressed ways of behaving which are nevertheless understood and complied with, often quite unconsciously. This culture or way of life in the factory has to be learnt by those who enter it. Persons who behave in ways contrary to it appear as maladjusted or deviant members of the factory. It is the culture of a factory that determines what ways of behaviour are open to individuals and groups in their relations in the factory[1].

In our discussions of attitudes to technological change, we have followed the theory developed by Newcomb (1952). The way events, persons, or things are perceived by different individuals depends upon their particular viewpoint or *frame of reference*. This frame of reference determines how these objects will be interpreted. The particular viewpoints or frames of reference which are used by an individual are largely determined by his past experience and by other individuals with whom he has had contact. In fact, successful communication between individuals implies that they are perceiving things in the same way, or in other words, that they are using common frames of reference. A person's *attitude* to an object is his predisposition to act, perceive, think, and feel with respect to it; it is related to a frame of reference. An attitude has an abiding and consistent character about it. Attitudes and frames of reference are related through their common perceptual characteristic, and they are interdependent in that they influence each other.

For all individuals certain frames of reference are dominant and these are used in the perception of quite different objects. Such non-discriminatory perceptions lead to generalized types of evaluation such as appear in racial and other prejudices.

An individual's attitude and his frame of reference are important factors in determining how he will behave in situations involving these objects. In this sense, neither attitudes nor frames of reference can be directly observed, but they can be inferred from a study of behaviour in appropriate situations.

[1] We are indebted to the studies from the Tavistock Institute for our particular usage of the terms social structure, role, status, prestige, responsibility, and culture, see Jaques (1951) and Trist and Bamforth (1951).

One form of such behaviour is an individual's expressed *opinion* about the object. This opinion may or may not accurately express the underlying attitude. The way a person behaves with respect to an object often belies what he has said about it. An opinion, formally or informally expressed, is only one of the cues which are necessary for the inference of attitudes and frames of reference. In the present study we were able to observe the behaviour of individuals in actual situations and this served to supplement the opinions they expressed.

One of the characteristics of groups of individuals is their tendency to share attitudes and common frames of reference. This sharing leads in its turn to common action by the members of the group. A concept which refers to this common sharing of attitudes and action is *group cohesiveness* as it has been developed by Festinger (Cartwright & Zander, 1954). A high degree of cohesiveness in a group is manifested in a sense of belonging on the part of the members which acts on them as a pressure to remain in the group. Depending on the cohesiveness of the group, the members share certain fixed modes of behaviour. The more cohesive the group, the greater will be the sanctions applied to deviants from the group. This concept has been largely developed for informal groups that can be characterized as peer groups. Such informal groups do exist in factories, but there are many other groups which find formal expression in the social structure. For example, there are groups of individuals who occupy similar roles, or belong to the same union, or are members of the Works Council. These formal groups are inherent in an hierarchical social structure like a factory or industrial company. An attempt to extend the usefulness of the concept of group cohesiveness for such formal groups is made in this book. The essential difference between the formal and informal groups is the extent to which elements outside the group are involved. The informal groups throw up their own leaders, largely determine their own behavioural situations, and exist because the individuals came together to form the groups. On the other hand, the formal groups have leaders who are often in a role which is different from that of the group members, are faced with externally determined situations, and exist because the members were put together by influences again originating from outside the group.

The informal groups are similar to the groups that have been used in many laboratory experiments in social psychology. Such peer groups do not provide directly useful information about many features of the groups in a structured system like a factory. In particular, they do not exhibit the interrelatedness of the groups and individuals in such systems.

INNOVATION AND THE PATTERN OF A FACTORY

In the dynamic pattern of a factory that has been described, change is continuously occurring. Changes in the social structure, the culture, or the personnel are all liable to occur as the factory continues its ongoing life. A change in any of these elements will initiate changes in the others which may maintain the original pattern or produce a new, more appropriate pattern. This ongoing development is essential to the life and progress of industrial communities. To understand the delicate interplay of the various influences in the life of a company is at once a problem of prime practical and theoretical interest. It is of practical importance because so often these changes are not achieved without considerable human unhappiness, however this is expressed in terms of economics, low productivity, social status, individual health, or broken human relations. Some understanding may lessen these effects in future situations. Theoretically, it is important if the role of psychological and social influences in human behaviour in real situations is to be understood.

Social systems undergoing rapid and specific changes are particularly suitable situations in which to study these dynamics. In such cases, the interacting factors become exposed to observation. The quasi-equilibrium state is upset and the unbalanced influences are to be seen in isolation. Existing positions in the social structure are altered suddenly and the individuals and groups in them are unable to readjust with the same speed. They rely on old frames of reference and attitudes to cope with the new situations. This often leads to inappropriate behaviour, which again is easily observed. As the individuals and groups struggle to maintain their existence, influences within them that normally are implicit and unseen become explicit and overt. For these reasons, we chose a company undergoing rapid and extensive technical change for our study.

REFERENCES

ARGYLE, M. *et al.* (1958). Supervisory methods related to productivity, absenteeism, and labour turnover. *Hum. Relat.*, vol. XI, pp. 23–40.

BAVELAS, A. in LEWIN, K. (1952). *Field theory in social science.* London: Tavistock Publications. pp. 230–1.

BLAU, P. M. (1955). *The dynamics of bureaucracy.* Chicago, Ill.: University of Chicago Press.

CARTWRIGHT, D. & ZANDER, A., (eds.) (1954). *Group dynamics.* Evanston, Ill.: Row, Peterson; London: Tavistock Publications.

CURLE, A. (1949). A theoretical approach to action research. *Hum. Relat.*, vol. II, pp. 269–80.

FRENCH, J. R. P. & COCH, L. (1948). Overcoming resistance to change. *Hum. Relat.*, vol. I, pp. 512–32.

FRENCH, J. R. P., ISRAEL, J. & ÅS, D. (1960). An experiment on participation in a Norwegian factory. *Hum. Relat.*, vol. XIII, pp. 3–20.

HOMANS, G. C. (1953). Status among clerical workers. *Hum. Organization*, vol. 12, pp. 5–10.

JACKSON, J. M. (1953). The effect of changing the leadership of small work groups. *Hum. Relat.*, vol. VI, pp. 25–44.

JAQUES, E. (1951). *The changing culture of a factory.* London: Tavistock Publications.

KATZ, D. *et al.* (1951). *Productivity, supervision and morale among railroad workers.* Ann Arbor, Michigan: University of Michigan.

LEWIN, K. (1952). *Field theory in social science*. London: Tavistock Publications.

NEWCOMB, T. M. (1952). *Social psychology*. London: Tavistock Publications.

ROETHLISBERGER, F. J. & DICKSON, W. J. (1939). *Management and the worker*. Cambridge, Mass.: Harvard University Press.

RONKEN, H. O. & LAWRENCE, P. R. (1952). *Administering changes*. Boston, Mass.: Harvard Graduate School of Business Administration.

TRIST, E. L. & BAMFORTH, K. W. (1951). Some social and psychological consequences of the Longwall method of coal-getting. *Hum. Relat.*, vol. IV, pp. 3–38.

Works not specifically referred to in the text

JACOBSON, H. B. & ROUCEK, J. S. (eds.) (1959). *Automation and society*. New York: Philosophical Library.

LEACH, E. R. (1954). *Political systems of highland Burma*. London: Bell.

NADEL, S. F. (1956). *Theory of social structure*. London: Cohen & West.

POLITICAL AND ECONOMIC PLANNING (1957). *Three case studies in automation*. London: P.E.P.

TWO

The Textile Company

The textile company in which this study was carried out is one of long standing and of high reputation both within and without the industry. Some part of the company has been employed in weaving cloth in various mills in the country for about 140 years, and since the invention of man-made fibres at the beginning of this century it has been in the vanguard of this development.

The two of the company's mills with which this study is particularly concerned are those at two small towns, Radbourne and Debenham, in the same part of the country and a few miles apart. In both these towns, weaving has long been practised, and the company itself has a long tradition of activity in the area.

As a background to the changes which occurred later, we will consider in greater detail the history of the two mills beginning first with Radbourne Mill, the older of the two.

DEVELOPMENT OF RADBOURNE MILL

The mill at Radbourne was originally a silk weaving factory, and during the nineteenth century it specialized in crepe weaving which brought the district considerable prosperity. Existing records of the factory conditions at that time indicate that throughout its history the company has been progressive as far as the welfare of its employees was concerned. As early as 1846, organized welfare and educational opportunities were offered for the workers by the company, and in the 1850s a nursery school was started for the children of working mothers.

During the nineteenth century the control of the company was firmly in the hands of the family which founded it. Indeed, the main members of the family lived in the immediate area of Radbourne Mill, and there are records of large gatherings of the factory folk at the house of the head of the company on several holiday occasions. As far as can be seen from the records, the head of the company during this period was a typical nineteenth-century industrial patriarch. A small reminder of the old nineteenth-century working atmosphere still exists in the mill at Radbourne today. This is in the form of several moral directives painted on the iron beams such as 'weave truth with trust'. These have been re-painted on every successive re-decoration.

In 1873 there were 1,337 people employed in the mill, and of these 1,223 were women. The number of looms reached 500 by 1884, but this figure had doubled by 1900. The beginning of the twentieth century saw an end to expansion of this type, since the rural population of the surrounding area was declining. Production continued to increase by means of improved technology, but the personnel employed in the Mill declined in numbers.

The early years of the new century saw the development of artificial silk, and over the years the company has woven more and more of this yarn and less and less real silk. Today, man-made fibres of all sorts are almost exclusively used in the weaving at Radbourne Mill.

From the time of the initial introduction of artificial fibres, the founding family gradually ceased to control the affairs of the company. At the present time, only one member of the immediate family is on the board of the company. This change has occurred during a period of rapid expansion of the firm as demand for the new fibres increased.

During the 1930s, the mill personnel worked a great deal of short time, but there were no dismissals, and employment by the company did mean a security that was lacking in many other parts of the country. This period has left its mark on the social pattern of the mill. First, the company is regarded as a 'good' employer by the older people in the mill. Second, a pattern of leisurely working was established which was not necessarily suitable to meet the requirements of modern automatic production.

In 1938–39 the company purchased 350 modern automatic looms from the United States, and although the majority of these looms were installed in another weaving mill, about 94 were put into service at Radbourne. At this period, less than 3 per cent of the power looms in Great Britain were automatic and the percentage weaving man-made fibres was even smaller. These modern looms were initially introduced in order to meet the growing competition of overseas manufacturers.

Just prior to the introduction of automatic looms which defines our present study, the executive management gave three reasons for installing them:

> 'First, economic. To stay competitive in this country, it is important to lower labour costs per loom and labour costs in the overall production. Then, the company has, as a whole, for its own benefit and for that of the national textile industry adopted a policy of automatizing. This is for competition internationally, rather than domestically.
>
> 'Second, quality. This is better from automatic looms for the types of cloth we want to produce and this is necessary for competition also.
>
> 'Third, labour. After the war, with the Beveridge plan, there was going to be a nation-wide shortage of labour and this was the sort of policy which was needed to meet this situation.'

The company records showed that extensive experiments had been carried out with various types of automatic loom, particularly a British automatic type. From these tests, the company had concluded that there was no British loom to compare with those produced in Switzerland or the U.S.A. for the weaving of man-made fibres. For the best use of yarn, they further concluded that automatic looms were preferable on both a quality and a cost basis. For example, the company estimated that shed labour costs per yard of cloth were in the ratio 1·1 : 1·7 : 2·2 for double-shift automatic weaving, single-shift non-automatic weaving, and double-shift non-automatic weaving respectively. The policy of introducing automatic looms was therefore taken up again after the war with the introduction of 264 Crompton & Knowles automatic looms at Debenham Mill in 1947–50.

With the introduction of automatic looms has come the need to work machinery for as long as possible in order to achieve economic production costs and obtain a faster return on the high initial capital cost. This has meant the introduction of shift work in the form of a double-day working shift on a five-day week excluding weekends. Shift working slowly spread from 1948 onwards to several sections of Radbourne Mill in both the weaving and non-weaving departments. This type of working was completely novel for most of the small group of workpeople concerned.

The change immediately preceding the one we are considering was the purchase in 1952 of 48 looms of the type it was planned to install in the major innovation of 1954–55. These looms were sited very near to the old automatic looms which had been installed just before the last war.

DEVELOPMENT OF DEBENHAM MILL

Debenham has a long textile history and is a centre in which the present company has been working for many years. In spite of this, weaving was not carried on in the company's mill there until about 1926. At that time, about 400 non-automatic French looms were installed as an extension of Radbourne Mill and the activities of the mill were then confined solely to weaving. Preparation of the yarn prior to the weaving process was performed at Radbourne and the prepared materials brought over to Debenham.

However, not long after the installation of the looms there was a recession in trade, and by 1933 there were only 90 looms in operation, each weaver having two looms. During the succeeding years trade improved, and by 1939 234 looms were running, all with certain modifications and additions which enabled them to run in a more satisfactory fashion so that by this time the number of looms per weaver had increased to about eight.

These conditions were maintained during the war, and then from 1947 to 1950 264 automatic looms were introduced into the mill displacing the older looms. The looms so displaced were sent to Radbourne Mill and some of these were only finally replaced by the automatic innovation which we studied. The automatic looms installed at Debenham are the same type of weft-replenishing looms as those now installed at Radbourne.

In addition to the introduction of modern, automatic machinery, Debenham began to take over some of the processes associated with weaving which had previously been done exclusively at Radbourne. From about 1945 onwards, gradual innovations were made, so that when the automatic loom installation was completed the mill was autonomous, from the introduction of the yarn to the removal of the woven cloth for the finishing processes.

Since 1945, there have been several other major changes which concerned the workpeople at Debenham. First, double-day shift working was introduced in 1947, and this enabled the new automatic looms to be run under this system. Second, the policy concerning the recruitment of weavers had changed. Previously, all weavers had been women, but now men formed the majority of such recruits in the mill, although there were still some women recruited as well. Lastly, and resulting from the other two changes, a special training school was set up at Debenham. The function of this school was to train new weavers and also new overlookers (loom mechanics), both because of the policy of taking on untrained personnel, and also because of the considerable increase in personnel requirements which had been made necessary by shift work. This school also trained some Radbourne operatives. The demands upon the training school were, of course, eventually met as far as Debenham was concerned, and in 1954 the school was closed.

Also during this post-war period, a controlled humidification plant was erected, and this aided work on the yarn at all stages of production, although the plant did not include a refrigeration plant for use in the summer months.

A minor trade recession in textiles in 1952 led the company to dismiss some employees and to revert to day working; recession conditions lasted approximately six months. Since then Debenham has been operating under fairly stable conditions as a unit of 264 automatic looms, with a consistent level of efficiency.

The total staff in October 1954 was 143, consisting of one manager, eight supervisors, and 134 operatives.

THE TRADE UNIONS IN THE MILLS

The two main unions are the National Association of Power Loom Overlookers, which caters for the loom mechanics, and the Transport and General Workers Union, which caters both for weavers and for many other workpeople in both mills. There are also the specialized maintenance personnel who belong to such unions as the Amalgamated Engineering Union and the Electrical Trades Union.

By agreement with the management, all loom mechanics must belong to the N.A.P.L.O. and, in addition, the number of loom mechanic apprentices in either mill at any time is a matter for negotiation between union and management. This group of workpeople is in a position of considerable strength and status, which is heightened by the fact that the loom mechanic is reckoned by the management to be the key technician in the weaving process. An indication of the strength of this group is their refusal to operate under a work-study scheme, although most other operatives in both mills do in fact work under such schemes and the implementation of work-study techniques throughout the mills is official company policy. Membership of the other main union, the T.G.W.U., is not 100 per cent. At Debenham it was about 90 per cent and about 70 per cent at Radbourne.

Unions were accepted by the company, and all negotiations concerning conditions of work and rates of pay were carried out between management and union representatives. On the whole, it is accurate to describe official union–management relations as cordial, and the disputes which occurred were generally solved by local negotiation at each mill.

ADMINISTRATIVE STRUCTURE OF RADBOURNE AND DEBENHAM

Each of the two mills is controlled by a manager with full administrative power for the organization and running of the mill to secure the final end of production – the manufacture of woven cloth. Other functions of the company are handled centrally – sales, yarn supply, cloth finishing, cloth storage, and so on.

Although there is considerable consultation between the mills and the company headquarters, nevertheless the placement of orders is in the last resort a company responsibility. A company programming office handles the main aspects of production control, including delivery of yarn to the mills, and the supply of woven goods to the customer as and when they are required. This and other company departments are housed in the same town within a few miles of the mills, with the exception of the sales department, which is located in a large city, and is tied to the organization via the headquarters administration. There is very little direct contact of any sort between the sales department and the mills, although sales trainees, in common with other company trainees, spend part of their training time at the mills.

In each of the mills, there is also a system of programming and control for working in conjunction with the central programming office, on the one hand, and the production staff, on the other. Naturally enough, this local system is more complex at Radbourne than at Debenham because of the greater size of the mill and the greater complexity of its production task.

This integration of production is very different from former days, when the mills had greater autonomy over the organization of production and so on. For example, they ordered and obtained their own supplies of yarn direct from the manufacturers. Present-day integration is via the company headquarters and contact between the mills is mostly of an informal nature, such as the exchange of advice, and the lending and borrowing of spare parts and materials.

TEXTILE WEAVING IN GENERAL

How do Debenham and Radbourne Mills fit into the picture of textile weaving throughout the country? A few facts and figures about the weaving side of the textile industry will help the reader to fit the mills into the broader pattern.

In May 1957, the total number of looms weaving man-made fibres, cotton, and mixtures of yarn was just over 300,000. Of these looms, nearly 16 per cent were automatic. Similar figures for 1948 revealed a total loom complement of just over 400,000, of which only 7 per cent were automatic. Although the decreased loom strength of the industry accounts for part of the increased percentage of automatic looms, in fact the actual number of automatic looms increased in the nine years from 28,000 to 47,000.

In 1957, there were 925 firms weaving man-made fibres and/or cotton, and the mean number of looms per firm was 326. This figure also gives a rough indication of the mean number of looms per mill.

In our case in 1956, the total number of looms at Radbourne and Debenham Mills was 688 and 264 respectively. This information indicates how the position at Radbourne and Debenham relates to the industry as a whole.

PROCESSES OF WEAVING

We have now, as it were, set the scene. But before we can describe the action, it is important to give the reader an outline of the techniques involved in producing woven cloth, and also of the work-study system that operated in the mill.

Broadly speaking, there are two component sets of processes in turning yarn into cloth. First, there is the weaving process itself, which is performed on a loom. The essential function of the loom is to intertwine lengthwise and crosswise threads of yarn in a systematic fashion in order to produce cloth. The loom achieves this end by passing the crosswise or *weft* thread over and under the lengthwise or *warp* threads and by pushing this interwoven thread of yarn firmly into place. Secondly, there are the preparatory processes, in which the yarn is assembled and arranged for weaving on the loom. Here again there are two main components; the preparation of the warp (*warping*) and the preparation of the weft (*pirning*). Warping itself takes place in three distinct stages; warping proper, *sizing*, and finally either *entering* or *knotting*.

We will now describe each process in a little more detail, beginning with the preparatory processes which precede the weaving process.

1. *The Warp*

Warping. There are various methods of making warps, but they all entail winding a series of single lengthwise threads of yarn off the containers on which the yarn arrives in the mill onto a wooden or metal cylinder known as a *beam*. Each beam is about six feet wide. The operative performing this process is called a *warper*, and normally one warper works one warping machine, except with some more modern warping techniques, where two or three warpers form a team to work on one machine.

Each thread of the warp is generally several hundred yards long, depending upon the length of cloth to be made, and each warp will contain a large number of threads side by side. In weaving man-made fibres, where the thickness of each thread is generally small, there may be as many as 10,000 *ends* of thread on one beam. The number of ends varies with the *density* and thickness of the warp threads in the finished cloth.

Sizing. When the warp has been made, generally speaking it is then passed on to the sizing department to be sized. In the sizing operation, the yarn is wound off the warping beam, passed through a bath of size, dried, and then wound on to another beam which will be put into the loom. These operations are all performed on a single machine which is supervised and controlled by two *sizers.*

The purpose of this process is to enable each thread of yarn to be given a protective coating of size so that the thread will not be damaged by the action of the loom mechanism. In later finishing processes, the size is removed from the woven cloth.

Entering. The sized warp next passes on to be entered. In this process, the warp threads are pulled through a framework known as a *shaft.* When it is in position in the loom, the shaft is raised and lowered vertically at right angles to the direction of the warp threads. By the use of several shafts, each containing only a certain number of the warp threads, some of these threads can be drawn up and separated from the others, thus enabling the weft thread to be inserted in the gap between the warp threads. The simplest weave needs only two shafts which pull up alternate warp threads so that the weft thread crosses each warp thread in an under–over fashion.

Variations on this simple pattern are, however, innumerable, so that is is possible to have a large number of shafts to pull up first one set of threads and then another, so forming whatever pattern of threads is required. Entering is performed almost completely manually. Operatives work in teams of two to prepare the shafts, and the more skilled of the two is known as an *enterer.*

When the entering is finished, the warp complete with the shafts and some other ancillary equipment is sent to the loom in the *weaving shed.*

Sometimes, however, the sized warp is prepared to be woven in a loom in which a similar warp was previously woven. In this case entering is unnecessary, as it is possible to use the set of shafts already in place in the loom. A few yards of the warp threads of the previous warp are left in the loom when the woven cloth is cut off. Then, all that has to be done is to tie or knot each thread of the new warp onto the appropriate thread of the end of the previous warp. The new warp threads can then be drawn through the shafts *in situ* in the loom, saving the time and trouble involved in assembling a new set of shafts. This process cannot be continued indefinitely, and after a certain number of similar warps have been woven in the same loom, the set of shafts has to be replaced for cleaning and repairing.

This *knotting* process is carried out by a *knotting machine*, and each machine is set up and controlled by a *knotter* and an assistant.

2. *The Weft*

The other 'half' of the cloth is made up of the weft (crosswise) threads. Preparation of the weft is simpler, in that there are far fewer component processes. The main process is winding the yarn off the original container and onto a *pirn* or *spool*. This is a short wooden cylinder about six inches long and half an inch in diameter, with a wooden flange at one end. Each pirn will hold a certain length of yarn, which naturally varies with the thickness of the yarn, but the great difference between a beam and a pirn is that the pirn holds only a small quantity of yarn, and so needs frequent replenishment.

When they have been wound, a number of completed pirns are slipped over steel pins fixed on a board and these boards are then taken and placed on the appropriate loom. The operatives who prepare the pirns are known as *spoolers*, and each spooler supervises the preparation of between 20 and 100 pirns, varying with the thickness of the yarn to be spooled.

To enable weft yarn to be woven into the cloth, the pirn is placed, by the weaver, in a *shuttle*, which is a rectangular wooden container into which the pirn just fits. The shuttle carries the yarn, which is unwound off the pirn through an opening in the shuttle, back and forth between the warp threads to form the cloth.

3. *Weaving*

We have almost covered the weaving process by describing the processes which are preparatory to it. However, a little more remains to be explained.

As we have said, a loom is essentially a mechanism for combining the lengthwise warp threads with the crosswise weft threads by passing the weft under and over the warp in a systematic fashion. A single passage of the shuttle across the loom, depositing one weft thread, is known as a *pick*.

Obviously, it is possible to weave two or more different types of weft yarn in with the same warp and in this case the loom contains *boxes* at the side of the loom in which the shuttles containing the different yarns rest when not travelling across the loom.

The basic operation of any loom consists of three steps:

(i) Movement of the shaft to raise some warp threads above the others.
(ii) Passage of the shuttle through the opening or *gate* which is formed by (i), leaving a trail of weft thread behind it.
(iii) Pushing the resulting weft thread firmly into place next to the preceding thread.

A *weaver* is responsible for looking after one or several looms, and his total number of looms is known as a *sett*. The number of looms in a sett varies from two to thirty or more. The weaver is not responsible for maintaining the looms in good working order, or for correcting loom faults other than minor ones. This is the task of the *overlooker* or loom mechanic, who also has a sett of looms which is generally larger than a weaver's sett, and so one overlooker commonly maintains the looms of two or more weavers. The overlooker is also responsible for preparing the loom for weaving when the loom has *felled out* (woven the warp right out). As we have previously pointed out, the warp is often several hundred yards long, and generally takes some weeks to be completely woven. The insertion of a new warp and the removal of the old empty beam is therefore a major turn-round job which takes a number of hours to complete. The completed roll of woven cloth, wound on a wooden beam, is called a *packet*.

Although there are other, less important ancillary tasks to both preparation and weaving, the account which we have given should enable the reader to understand the broad processes as they are referred to in later chapters.

CRITERIA OF PRODUCTIVITY

Before we consider the work-study scheme, we must briefly discuss the terms *efficiency* and *damage*. These two factors are the overall criteria by which the productivity of the looms is gauged. Efficiency is the percentage of work achieved compared with the maximum possible work at a given loom speed. The figure takes into account factors beyond the control of the factory, such as yarn shortages or power failures.

Damages are all those characteristics of the woven cloth which make it commercially sub-standard. There are many different types of damage, but the important ones are those which persist throughout large lengths of the cloth and thus make it into what is called a *damaged packet*. Damages are very difficult to evaluate quantitatively – unlike loom efficiencies. Minor, non-persistent damages can obviously be quite severe, but they are not disastrous since the particular piece of cloth can simply be cut off and destroyed.

THE WORK-STUDY SCHEME

As this scheme is intimately connected with some of the consequences of introducing automatic looms, a short account will enable the reader to follow more clearly the details in following chapters.

The work-study schemes were originally introduced as an overall company policy shortly after the war, and were the subject of consultation between the unions and the management. During our period at the mill, there were often occasions on which unions and management negotiated over alterations in the scheme.

Basically, the work-study engineers endeavour to fix a reasonable work complement or *load*, by carefully analysing the component operations of any particular task and then timing such operations over a large number of hours. On the basis of the work-load, an incentive bonus scheme has been constructed to pay bonus money over and above the *base-wage*. Theoretically, the work-study schemes were so designed that an ordinary worker could expect to earn a fairly regular percentage of his total earnings as bonus money, providing that his load was steady. Bonus money plus base-wage is known as the *target wage*.

Quite clearly, many things could affect the work-load, e.g. differing thicknesses of yarn, the speed of the machine, difficulty of processing different types of yarn, and so on. This meant that the work-study department was constantly busy checking previous measurements and measuring new work as conditions changed. A worker's load was made up by adding together all the *factors* for the varying types of work which were included in his total task. An example will clarify these points: a weaver has 20 identical looms, all weaving the same cloth. Each cloth (or *sort*) has a *sort factor* which comprises standard times for the work involved in weaving that particular cloth on one loom. This includes mending breaks in the yarn, changing over shuttles, supervising the loom, and so on. We will assume that the sort factor for this particular cloth is 5. Then with 20 looms the weaver has a total load of 100, although there are certain additions, e.g. walking time, which are also taken into account. A load of 100 factor points is the optimum load on which the bonus scheme is calculated. Naturally, as different types of cloth are woven, the factors will vary and so the actual load will sometimes be above and sometime below the ideal figure of 100.

Without any more detail, we have now outlined the basic aspects of the scheme. Each worker has a work-load which is ideally 100. This is composed of factors for the total number of processes for which the worker is responsible, and each factor contains elements which cover the active work operations such as stopping a machine, repairing damage, supervising a machine, walking to the machine, and so on.

The reader is now in possession of the basic vocabulary of the mills which we studied. In the chapters that follow some of the social and psychological implications of these terms will emerge, and the technology of weaving will be invested with some of its human meaning.

THREE

Debenham Mill

In this study, we have in many ways used Debenham Mill as a yardstick by which to measure the changes which occurred at Radbourne. This resembles the use of a control against which changes can be evaluated, although in this case we were not able to equate the variables between the two mills – naturally enough! Therefore, the analogy of the yardstick is probably a more accurate one than the analogy of an experimental situation with a 'control'.

Our reasons for considering Debenham Mill as part of our research situation were really twofold. First was the fact that the mill at Debenham was operating the same type of automatic loom as was to be introduced at Radbourne, and operating it very efficiently. This mill was, as it were, a pure culture of the operation of this particular type of automatic loom. Radbourne was, from this point of view, impure because of the presence of other types of loom, but we felt that there was a good deal of similarity which would be of great value to us.

Second, Debenham represented a stable system as far as the functioning of automatic looms was concerned. We felt that, whatever the problems which current production might bring, the overall problems of running an automatic unit had been largely dealt with, and the existing system represented a completed process of adjustment to automatic production.

For these reasons, therefore, we have in various parts of this book compared the two mills, trying by reference to the one to highlight the changes currently occurring in the other. *Figure 1* shows the layout of Debenham Mill.

FIG. 1: *Layout of Debenham Mill*

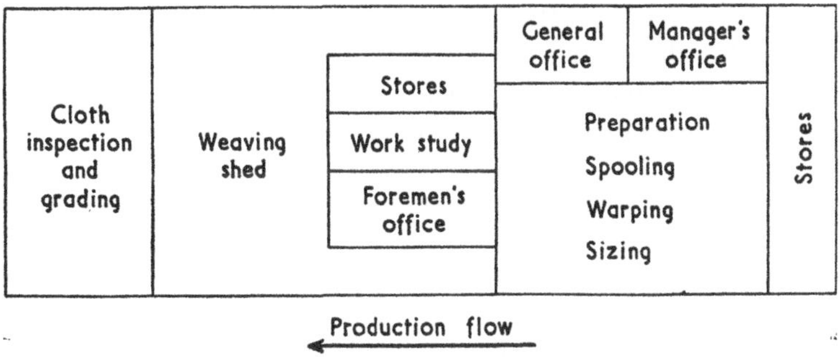

Cloth inspection and grading
Weaving shed
Stores
Work study
Foremen's office
General office
Manager's office
Preparation
Spooling
Warping
Sizing
Stores
Production flow

There are some points which the plan does not reveal. The mill is only single-storied, although certain offices have been built in specially constructed half-stories. Its total size is 58,300 square feet. Top-lighting is used throughout the building, which is constructed of bricks. The first general impression on entering the building is of a light, clean, and airy working environment.

The building itself is on a larger site which also contains a multi-storied building concerned with the preliminary processing of yarn before it is sent to the weaving mills. This building houses the main services for both units, and also the maintenance engineering staff. Each building, however, is under the separate control of a manager, both managers being subordinate to the central executive.

What the plan *does* reveal is the 'straight-through' aspect of the processing. Although there is nothing resembling a production line, nevertheless the product moves in the same general direction as it passes from stage to stage, and thus the yarn is brought in at one end of the building, and the woven cloth is taken out at the other end.

The mill is sited, with some other factories of different kinds, just outside the centre of the small town. A large majority of the workpeople are drawn from the local area, and many of them cycle to and from work each day.

ORGANIZATION OF THE MILL

The mill's overall function is to produce woven cloth from yarn. To do this, the overall task is differentiated in the way shown in *Figure 2*. It should be noted that, at Debenham particularly, the terms *supervisor* and *foreman* were used synonymously.

FIG. 2: *General Organization of Debenham Mill*

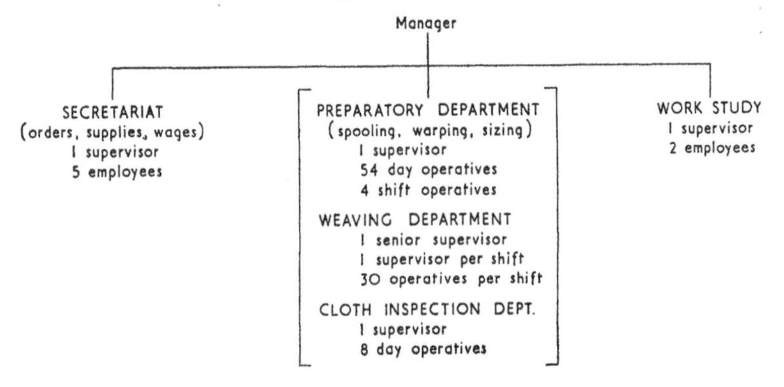

Inspection of *Figure 1* on p. 27 will show that, in the main, the physical layout of the production part of the mill also falls into three main sections corresponding to the differentiation of the work function. Thus the physical structure and work function are very closely linked. Since the service departments of work study and secretariat serve all departments, their siting is not perhaps so relevant.

Some account of the development of Debenham has already been given in Chapter Two, and the reader will recall that the weaving mill had been established since 1926, although the preparation of yarn and inspection of cloth were not carried out at Debenham in the first place. After the war, in 1945, there were 234 non-automatic looms running. These looms had, however, been modernized to such a degree that they in many ways resembled automatic looms. However, they were not truly automatic looms.

Between 1945 and 1947 yarn preparation processes and cloth examination had been started, so that by the time the older looms began to be replaced by the modern American automatic looms from 1947 onwards, the mill already had the autonomy which it has today. Much of the equipment in the preparatory and inspection sections was modern in design and function, although at the time of our study there was also older machinery in use. The company was, however, planning to replace the oldest of this machinery with new automatic machines. Thus the physical plant and the organizational structure can be said to facilitate the smooth flow of work which automatic techniques of all types demand for efficient operation.

The new automatic looms were installed gradually as they became available, and when they had all been installed and properly run in, the efficiency of the weaving shed as a whole stood at about 80–85 per cent.

However, after a recession in production which occurred in 1952, the loom efficiency began to rise steeply until the overall figure was between 90 and 94 per cent. Since that time, and during the period of our research there, this figure for loom working efficiency was maintained (see *Table 1*). The reasons for the increase in the efficiency of the looms were, unfortunately, not available to us since the events had been too overlaid with the passage of time.

A complicating factor during the installation of the looms at Debenham was that the company was at the same time introducing work study throughout its various units. Here, too, real information concerning possible difficulties which arose out of this policy was virtually impossible to obtain because there was much feeling attaching to the idea of work study and its application in the mill.

During the period of the research study, there was little interruption in the highly efficient production of cloth. Difficulties did, of course, arise. But these were difficulties stemming largely from external circumstances and not from internal ones. The

TABLE 1 WEEKLY LOOM EFFICIENCY, DEBENHAM MILL

Date	*Mean weekly efficiency %*
September 1954	92·3
October 1954	93·3
November 1954	93·2
December 1954	92·8
January 1955	92·7
February 1955	92·6
March 1955	91·7
April 1955	92·5
May 1955	93·4
June 1955	95·0
July 1955	94·8
August 1955	94·8
September 1955	95·0
October 1955	95·1

major internal difficulty was a re-allocation of looms stemming from changes in the nature of the cloth that was woven. However, after a short period of uncertainty among operatives and supervisors the re-allocation was completed successfully.

MANAGEMENT ORGANIZATION

Figure 3 shows the formal management organization of Debenham Mill. During the research, Anderson took another post in the company.

FIG. 3: *Formal Management Organization of Debenham Mill*

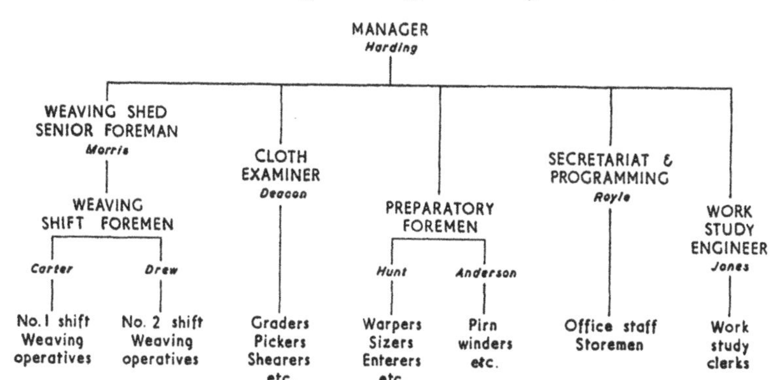

Although this is the recognized way of setting out the relative position of the management group, it does not do justice to the social operation of this group. Functionally, *Figure 2* indicates fairly closely the actual grouping within the larger group. There was naturally a much closer link between the weaving supervisors than between, say, the cloth room supervisor and the chief clerk. But more noticeable than this sub-grouping was the general cohesion of the group as evidenced by easy, informal relationships existing between virtually all members. In effect, the management group was composed of two levels only – Harding the manager, who was also group leader, and the rest of the supervisors under his control. The chart also shows a senior weaving supervisor, but this, though correct according to the formal structure, was not the case in procedure. Mr Morris took over the job of senior supervisor when the previous man, Laycock (now assistant manager at Radbourne Mill), left the mill to take a staff post. He did not, however, exactly succeed Laycock since Laycock was senior supervisor over *all* departments. Laycock's position was, therefore, more nearly that of under-manager.

The fact that his own post was different from his predecessors' was recognized by Morris who said in describing his main task: 'My job is primarily that of watching quality.'

This definition of his role was also supported by operatives – for example: 'I think Mr Morris is mainly concerned with the quality of the cloth.'

From observation, Morris was considered by the other supervisors as being one of themselves. He did not interfere with the running of the shifts so that the shift supervisors were largely autonomous – and both shift supervisors said that in difficult matters they would consult Morris for his advice and assistance.

In addition to the groups in the mill, the manager also had access to a number of specialist advisers at the company headquarters. These specialists had no executive power in the mill, and could only exercise authority through the senior management group of the company.

There was little formal contact between the two management groups at Debenham and Radbourne – although the mills were only a few miles apart. The managers of the two mills met at managers' meetings and also occasionally visited the other mill. This lack of contact was even more marked for the other supervisors who rarely met, except perhaps at functions arranged by the local Textile Society.

During the installation of the automatic looms at Radbourne, there was some limited contact between the mills for a few selected people, and this is described in greater detail in Chapter Four. In addition, of course, one very important connecting factor was the appointment of Laycock, the previous senior supervisor at Debenham, to the post of assistant manager at Radbourne.

PATTERN OF OPERATION

As we suggest in this book, and in keeping with Trist's concept of the socio-technical system (1951), a particular form of technology will give rise to a certain pattern of operations. This pattern will be largely determined by three factors:

(*a*) *The technique itself.* In our particular case, the differing number of automatic and non-automatic looms which one operative could handle is an example of this factor.

(*b*) *The culture or 'way of life'*. For example, until after the war it was part of the culture of Debenham and Radbourne Mills to employ women as weavers and not men.

(*c*) *The personality of the individuals concerned.*

For the organization to work effectively all these three factors must exist in harmony. Some tolerance for minor fluctuations usually exists, but beyond the limits of this tolerance change must occur in one or more factors if their harmonious relationship is to be restored and an adequate operating pattern maintained.

On the evidence of company executives, and of individuals in the factory at Debenham, the pattern of working there was harmonious and very satisfactory.

For example, the manager said: 'I like to think of the people in the mill as a family'.

And an operative who also held an official trade union position said: 'Relations are as good at Debenham as you could expect them to be anywhere'.

Thus the solutions to problems of technique arising from automatic looms were presumably in harmony with the other elements which we have described.

Probably the best description of the way in which Debenham worked is to say that it worked cooperatively. We felt that the key to this cooperation lay in the cooperation of the management group in the overall production tasks of the mill. In a previous section we have described how the management group was structured on only two levels and how relations between the two levels – manager and supervisors – were relatively easy and informal.

The keystone of this management cooperation was almost certainly a daily production meeting held more or less regularly by the manager in his office. The development of this important meeting is interesting in itself and so we propose to discuss it in outline here.

It is company policy that the supervisors in each mill shall meet regularly to discuss matters of interest and importance to themselves and the company. At Radbourne Mill and at Battersley Mill (the company's mill in the North of England), this meeting took place formally at monthly intervals, with a formal agenda, minutes, and an official secretary. At Debenham, however, the situation was different. There was no formal meeting, but rather the daily production meeting. The agenda was informal and determined by day-to-day problems, and no minutes were taken.

Some years ago the Debenham meeting was a formal monthly one as at the other mills. It was held in the canteen away from the main production departments. Then, because of external circumstances, it was decided to hold the monthly meeting in the manager's office. The manager was not keen on this idea, and finally he decided to hold an informal meeting each day, and this has continued since 1951 in this fashion.

In our initial observations at the mill it seemed to us that this practice was worthy of more detailed study and the manager kindly granted one of us permission to attend a series of these meetings. We recorded the number and type of the contributions which members of the group made, and from nine recorded meetings the following analysis emerged:

TABLE 2 INTERACTION AT DAILY MEETINGS, DEBENHAM MILL

Type of Interaction	*% of Total Contributions*
Manager to a specific supervisor	34
Specific supervisor to manager	33
Manager's general comments	11
Supervisor to supervisor	21

One of the most interesting points in this table is the record of one-fifth of the total comments being between supervisors. This did not mean anarchy within the meetings, for this was certainly not the case. But the members of the group obviously felt free to comment and question each other directly.

The ease with which criticisms and feelings generally might be expressed was recognized by the supervisors themselves.

> 'If you have anything to say about another department you can say it there without any worry about them thinking you went behind their backs.'
>
> 'It helps to reduce the differences which arise between preparation and weaving departments.'

Thus the management group was daily in action as a small, informal, task-oriented group, solving current production problems and both giving and receiving information from the manager.

The question now arises of how this pattern fitted in with the operation of the mill. The two main factors which arise from using automatic looms as opposed to using non-automatic looms are these:

(*a*) Continuity of production arising from the fact that, ideally, an automatic loom need never stop.

(*b*) Increased speed of production arising partly from an actual increase in loom running speed, and also from the continuity.

There are other factors, but these are important ones from the point of view of the pattern of working of the mill. The high costs associated with an automatic loom installation also increase the pressure for continuity of working.

In terms of operation, these requirements of the technology lead to two things:

(*a*) Pre-planning to ensure that no loss of production is caused by materials and men not being at the right place at the right time.

(*b*) Swift communication to ensure that delays due to errors or breakdowns are corrected with the least possible delay.

From both these viewpoints, the production meeting was obviously a considerable aid, not only because decisions could be taken there, but also because it almost certainly increased the cohesiveness of the supervisory group and tended to decrease the likelihood of communication barriers being erected between one department and another. Although there are important reasons for not drawing too close a parallel between laboratory studies and the industrial scene, Argyle (1957) quotes several studies which support this observation.

Consultation between supervisors did not stop at the production meeting. On two occasions during the research study, we surveyed the contacts taking place between members of the management group during a working week. On both occasions in 1955 and in 1956 each person in the group contacted every other person in the group during the course of the week. The rate of interaction between any two individuals naturally enough varied a good deal.

From the operative's point of view – as we shall show in Chapter Nine – under automatic weaving conditions his or her supervisor becomes a very important communication link between his own department and other departments which serve it. In no sense does the operative's participation in communication increase, but the link between operative and supervisor becomes increasingly important. This is essentially a two-way process, with the supervisor relying on cooperation from the operative to enable him to act when operatives report faults or blockages.

Therefore both at the planning stage and at the breakdown stage, cooperation between supervisors is vital in order to satisfy the two conditions of continuity and speed of operation of the looms, and both the organization within the management group and the comparative freedom and informality of the members appeared to present the right conditions for cooperation to develop.

EXTERNAL FACTORS IN THE PATTERN OF OPERATION

We have already described how the overall planning and control was centred in the administrative headquarters of the company. At one time, though each mill had not gone out to seek its own orders, it had nevertheless, been responsible for arranging for its own supply of yarn. Now, however, the location of an order in one particular mill or another, and also ordering the yarn was done by the appropriate sections at the headquarters of the company. Both the allocation of orders *and* a properly-timed supply of yarn could make a great deal of difference in designing a production programme. In allocation of orders, the mill was fortunate in that it did not have to deal with a large number of different types (or *sorts*) of cloth, differing in yarn, weaving pattern, and so on.

Within the limits of economic circumstances the aim of the headquarters office was to give fairly large orders to Debenham, which meant that most of the looms would be weaving the same sort of cloth and that the changes occurring between cloths would be infrequent. During our study at the two mills, Debenham handled approximately fifteen different types of cloth, and Radbourne approximately seventy to eighty. It is interesting to compare this difference with one of the findings of an English team from the rayon industry which toured the United States to report on practice there. They noted that a differing economic balance of supply and demand enabled the American mills to weave a cloth in many more looms for much longer periods than was the case in this country. This sort of situation obviously facilitates planning.

The second main external factor was the ordering and supply of yarn to the mill. This too was now a function of the headquarters office. During our study, however, Debenham suffered, as did Radbourne, from inadequacies in the supply of yarn so that on occasion the supply of yarn to the looms ran extremely short. The only remedies which lay in the province of the mill were either to complain to the headquarters office, or to borrow some yarn from the other mill – provided that they had any stock of it. During one stage of our study, shortage of yarn for automatic looms was acute at both mills and each loom was rationed. We noted that the automatic loom weaver was more concerned than his non-automatic counterpart to keep the loom pirn battery full of spools – which amounted to about a dozen pirns – and a full board of pirns in position on the loom ready to be put into the battery. In times of yarn shortage, this 'stock-piling' of pirns had to be discouraged, although this restriction was not popular with the weavers.

SUMMING UP

Automatic looms present two immediate problems arising from the nature of the looms:

1. More precise methods of preparing and handling the yarn since the looms are intolerant of faulty material.
2. An emphasis on the continuity of running of the machine with all that that implies in terms of the right things and people being present in the right place at the right time.

Both these broad problems of quality of work and quantity of work had apparently been solved at Debenham. Problems of better quality had been handled by up-grading the quality of the skill exercised by all the workpeople in the mill. Although, of course, we were unable to observe the tensions which may have arisen during the up-grading process, there was nothing present in the mill to show that the new standards had not been quite satisfactorily assimilated.

The quantity aspect of automatic loom production relied in essence on the simplicity of the communication and control system. This seemed to derive from the close integration of all parts of the mill, which was run – at least implicitly – as a co-operative enterprise. Decision-making and planning were carried out on a production basis by the whole management group, and decisions made by one supervisor were liable to correction or alteration by the group as a whole if they did not tie in with the overall target of production. In addition, lines of communication were short and therefore distortion was minimal.

The smooth running of the mill was further supported by the policy of the company executive management who attempted to ensure that the most suitable types of cloth in the longest possible runs were woven at Debenham Mill. Some factors, however, operating against efficient production were also attributable to external factors.

Finally, it is important to remember that the physical size and geography of the mill contributed considerably to its particular social climate and efficient functioning.

REFERENCES

ARGYLE, M. (1957). *The scientific study of social behaviour.* London: Methuen.

TRIST, E. L. & BAMFORTH, K. W. (1951). Some social and psychological consequences of the longwall method of coal-getting. *Hum. Relat.*, vol. IV, pp. 3–38.

FOUR

Radbourne Mill – The Coming of the Automatic Looms

A STABLE FACTORY?

The company introduced 112 automatic looms of the pirn change type into Radbourne Mill in the winter of 1954–55. The directors perceived this step as a major departure from the existing pattern of the factory, which they described as stable and unaccustomed to change.

An examination of the factory records since 1945 or even since 1950 revealed that a constant stream of changes had, in fact, occurred in equipment, organization, and personnel. Looms had been replaced by other looms and new machinery had appeared in the preparatory departments for weft and warp. New yarns had been developed which led to changes in the type of cloth woven. Many new operatives had joined the company as others left, and significantly, a small number of men had gradually been recruited and trained to become weavers – a role previously occupied solely by women. Most of the existing members of the management group had only reached that level in the organization during this period. Oakroyd, the manager, himself came to the factory in 1950 from Lancashire, where he had managed a non-automatic weaving mill. In November 1952, 48 automatic looms, the same as those of the present innovation, were introduced into the Mill.

Why, then, was the present introduction of 112 automatic looms seen and treated as a major change compared with which the previous changes seemed to form part of a relatively static pattern?

The answer to this question is in large measure the substance of this book; but an examination of the existing factory prior to the innovation indicates the reason for the directors' attitude.

The geographical layout of the factory in mid 1954 is shown in *Figure 4*.

FIG. 4: *Layout of Radbourne Mill in mid* 1954

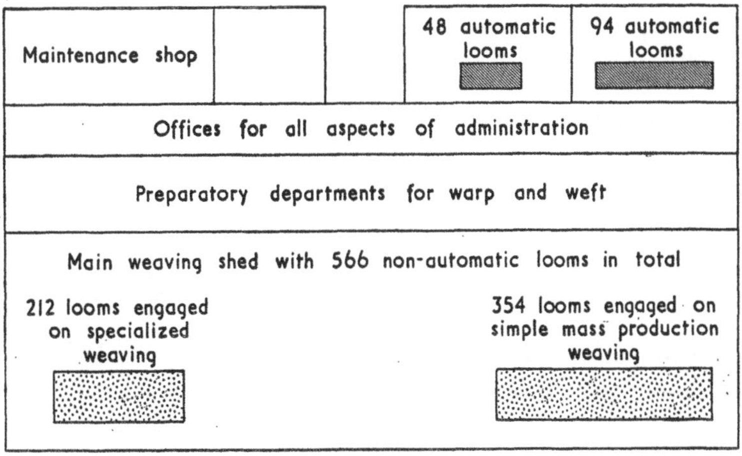

Maintenance shop

48 automatic looms

94 automatic looms

Offices for all aspects of administration

Preparatory departments for warp and weft

Main weaving shed with 566 non-automatic looms in total

212 looms engaged on specialized weaving

354 looms engaged on simple mass production weaving

The total area of the mill is 226,000 square feet. Two sections of automatic looms are indicated, both of which are situated apart from the main building of the mill.

The larger of these sections consisted of 94 automatic looms of the shuttle change variety which had been in the mill since 1938. The other section of 48 automatic looms, as mentioned earlier, arrived in 1952. Despite their existence for these considerable periods of time, both could be regarded as addenda geographically and, as we shall see, in other ways, to the main body and pattern of the mill.

A NON-AUTOMATIC MILL

The pattern of the mill was essentially a non-automatic one. The great majority of the weaving personnel were in the non-automatic departments. The preparatory departments spent most of their time on warps and wefts for the non-automatic looms. The shed of non-automatic looms was significantly referred to as the 'main' or 'big' shed. Most of the cloth (70 per cent) from the mill came from this shed. The management organization in June 1954 rested on the group of departmental foremen. It will be seen in later chapters that this feature was also part of the non-automatic pattern. It is shown diagrammatically in *Figure 5*.

FIG. 5: *Formal Management Organization of Radbourne Mill in mid* 1954

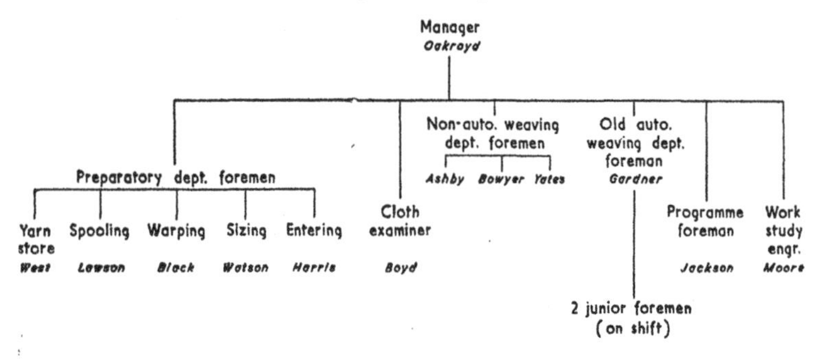

The automatic loom section existed detached from the non-automatic part of the factory and its presence did not materially affect the long-established pattern of the mill.

The introduction of 112 more automatic looms meant the beginning of a disruption of this balance. These looms, together with the existing 48 pirn change looms, were to be placed in the centre of the main shed, and non-automatic looms were to be removed to make room for them. A major part of the output would now come from the automatic looms; preparatory departments would be geared to their requirements. Automatic production would be seen as the ultimate trend for the mill. Non-automatic production would have received its death-sentence and only the unspecified time for the execution would remain. The full significance of the innovation for the pattern of structure, culture, and personnel of the factory was at this stage largely unknown. However, the above material changes were sufficient to explain the directors' attitude to the innovation.

CHANGES IN PERSONNEL AND SOCIAL STRUCTURE

Two other changes which occurred early in 1954 cannot be separated from the innovation of the automatic looms at the end of that year. The first involved the operatives. Six men were recruited specifically for training as weavers for the new automatic looms, and two overlookers – one experienced in automatic looms – were also engaged from Lancashire. In September 1954, the trainee weavers were in the small shed of 48 automatic looms, either helping the four existing experienced weavers or in two cases actually in charge of a sett of 16 looms. These four experienced weavers – two men and two women – were the survivors of a group of fifteen who had worked in this section since its inception in Novemeer 1952. Four men and four women had left the company and three other women were now employed in other sections, two as helpers and one as a non-automatic weaver. Two of the four weavers had been on these looms since they arrived and the other two for 16 months and 9 months respectively. Irwin and Young – trained at Debenham Mill – had been the overlookers for these 48 looms throughout and in 1954 they were joined by an apprentice overlooker.

The second change involved the management, that is, all the personnel above the position of chargehand. In July 1954, Laycock was appointed to the mill as assistant manager. He had been the senior foreman at Debenham Mill during the innovation period there, and had been sent to America by the company at that time to gain experience of the automatic looms from the manufacturers.

Thus, just prior to the innovation, the management organization had been modified from that shown in *Figure 2* (p. 28), by the insertion of a role between those of manager and departmental foremen.

In September 1954, the production personnel in the factory numbered 387 of whom 17 were management.

PLANS FOR INNOVATION

In August 1954, Oakroyd, the manager, made the first general announcement of the projected installation of automatic looms in the non-automatic shed to the Works Council and to the Foremen's Meeting. A few days later Manning, the company personnel manager, suggested that more information should be given directly to the personnel who would be involved in the change. He believed that this would help to invoke cooperation from the operatives in the changeover period. Oakroyd agreed to this suggestion and called two further meetings – a Foremen's Meeting and a meeting of all the operatives in the non-automatic weaving section from which looms were to be replaced. Manning addressed these meetings and emphasized the economic need for the change to automatic production. Only in this way could the company hope to remain competitive in the industry. He instanced the town of Lavenham – dependent, like Radbourne, for its prosperity on a single industry – which in the past failed to keep pace with industrial developments and lost all its wealth and most of its people. He said that changes would be necessary for some of the people, but that the company was seeking ways to make these most easily. A few questions about the details of shift working followed.

After these meetings the foreman of this non-automatic department obtained favourable responses from ten of his people about working on shifts. Oakroyd then drew up a list of sixteen weavers and five overlookers who, with the existing four weavers, two overlookers, and an apprentice, were the prospective operatives for the 160 automatic looms.

MACHINE MOVEMENT BEGINS

During October the 48 automatic looms were moved to new positions in the centre of the big shed and 32 non-automatic looms were removed and sold by the company. As soon as the installation was complete, warps were gated and the looms were started. Four weavers and two overlookers, now redundant in the non-automatic section, joined the automatic personnel as trainees. All these operatives were now paid on fixed rates which were calculated from their average bonus earnings immediately prior to the change. Management supervision of this automatic section in the main shed was jointly shared by Gardner, the automatic shed foreman, and Ashby, one of the non-automatic foremen. They now emerged as the occupants of a new position in the management structure. Their role was to supervise directly the new automatic section and to be senior foremen in the old automatic and non-automatic sections respectively. This was a planned intermediate step to the creation of two positions of shift managers when the new looms arrived.

THE HUMIDIFICATION PLANT

At the same time as these looms were being moved, the humidification plant in the main shed was dismantled in order to install a new automatic one. As originally planned this would have been completed before the looms were moved but delivery delays prevented this. For the first month of operation with the 48 automatic looms in their new position, there was no source of heating or humidity. Eight of the looms had a dress goods cloth which wove easily, but the remainder, weaving a satin cloth, were badly affected. Despite the fact that each experienced weaver had at least one trainee weaver with him on his sett, it was common to observe at least two looms in any group of eight stopped together. Hard frosts at this time added to the general discomfort and difficulty for the weavers and overlookers. The following were typical comments from a weaver and an overlooker:

> 'Things are awful, I've had the overlooker for an hour and a half and still the loom won't work. It's cold and the temperature was 58° yesterday. Why it's even cold to be working. I can't understand why the company do a summer job like this change in the winter. The organization is beyond me.'
>
> 'It's hopeless without the humidity. You can't be either orthodox or unorthodox. I could spend all my time on a couple of looms.'

The work-study engineer confirmed the reality of these remarks by the break frequency figures, and the weaving foreman quite accepted the poor output and bad quality, and said: 'Things were quite hopeless last week; yarn contracted and frayed, making breaks – the looms had to be continually adjusted to the changing conditions – even the leather varied with the temperature and the humidity'.

Oakroyd was absent from the mill during the second week of November, and in this week several weavers in the automatic section reacted against their conditions. Two of the trainee weavers – both living in villages some miles from Radbourne – gave notice and left the company, despite an offer from Laycock of a possible increase in their wage. They both gave the shift work as their reason for leaving. Another of the most promising of the trainees who had full responsibility for a sett of the looms in September also announced his intention of leaving because of the low fixed wage he was receiving compared with the other responsible weavers. Faced with this additional loss of a responsible weaver, Laycock investigated and found a general discontent among the weavers with their fixed wages in their present difficult conditions.

Further action by these weavers or the others was postponed until Oakroyd's return, after Laycock had expressed his sympathy with their complaint and made the promise of a wage review. Laycock had in fact urged Oakroyd for several weeks to raise the weavers' wage to retain them in this difficult transition period, but no decision had been reached before Oakroyd's departure. Laycock further assured the weavers that they were not held responsible for the poor output and that the humidification plant would be operating by the following Monday. The latter assurance was sceptically received by operatives and management alike. A weaving foreman remarked, 'No one but Laycock believes the plant will be ready on Monday.'

On Oakroyd's return the promised review was made, and all the trainee weavers had their wages raised to new levels relative to the four experienced weavers according to their past and present responsibility. With this arrangement, the weavers' notice of leaving was withdrawn. These rises lifted the wages of the weavers and trainees in the new automatic section above those of equivalent weavers in the other weaving departments of the mill. The foreman in the section remarked: 'We are trying to create by money that it is a good thing to be in the new section.'

From the beginning of November another 24 non-automatic looms were stopped as they felled out, releasing two weavers and an apprentice overlooker who joined the automatic section as further trainees.

The first batch of the new looms was expected early in December but in mid November Oakroyd told the Works Council that they had been delayed three weeks in the U.S.A.

The humidification plant did begin operation at the end of the third week in November – three days after Laycock's prediction but still earlier than most people expected – although it was under manual control owing to a separate delay in the delivery of the automatic switches. There was an almost immediate improvement in the weaving in the automatic section and an appreciative response from the operatives who commented:

> 'It's nice to feel warm again.'
>
> 'Things are better now – temperature and humidity are important to the looms, especially to the leather strap around the picking stick, which is the most vital part of the loom.'

The efficiences of the looms during this period, together with the efficiences for the whole period of the study, are given in *Figure 6*, p. 62.

BONUS PRODUCTION AGAIN

Bonus earning began again in the section in the first week of December. The four experienced weavers and two others were held responsible for all the cloth from a full sett of the looms and were paid according to the efficiency – that is the quantity of cloth from the sett. The trainee weaver remained on fixed wages.

The managers and the weaving foremen decided that the low efficiences could no longer be accounted for by the general conditions and settling-in of the looms. This opinion was supported by the experienced overlookers: 'It's the satin yarn itself now that is the limiting factor – not the conditions or the looms.'

The weavers unanimously held the opinion that the looms had not settled in, but all of them increased their wage by at least 4 per cent in this first bonus week. This change in the wage conditions of some of the weavers brought immediate complaints from some of the trainee weavers on fixed wages: 'It's not fair – we do just as much work as the others. We've been here too long waiting for the new looms.'

On the other hand, the responsible weavers began to complain of the trainee weavers for not pulling their weight. As the efficiencies rose, so also did the damage rate, which was interpreted variously by the management as due to the speed of the looms or to the approach of Christmas. It was not related directly to the contemporary change from fixed rate to bonus conditions. In an attempt to investigate the effect of speed on the quality of the cloth, the management did lower the speed of some of the looms by 5 per cent.

The experiment with the speed of the looms proved inconclusive and the speeds were raised to their former level. The lower speeds had meant a loss in production, to which the downtime when warps were changed also contributed. During December the downtime varied from 7 hours to 28 hours.

Absenteeism in the non-automatic weaving section throughout this month and January made necessary the temporary withdrawal of some of the trainee weavers and overlookers from the automatic section.

MORE CHANGES IN THE STRUCTURE

On 29 December, Oakroyd announced the appointment of Gardner and Ashby as shift managers. The evolution of these new positions in the structure was now complete. The new role carried responsibility for the whole mill during the hours outside the day work hours and involved a share in the interdepartmental planning and coordination, especially in relation to the weaving departments on shifts. Nevertheless, as far as the new automatic section was concerned the occupiers of these positions, despite their new title, remained the only management representatives between the operatives and the assistant manager.

In January more non-automatic looms were organized for shift work, and this delayed the transfer of an overlooker who was to have joined the automatic section for training at this time.

By this time the new looms had arrived in England but were delayed by Customs formalities.

A new *back beaming* warping machine, which was to produce beams for a fast sizing process known as *slasher sizing* (essentially for the new automatic looms), was installed in what had been storage space in the entering department.

THE HOOTER CAMPAIGN – AN AUTOMATIC WAY OF WORK?

With the new year the management initiated a campaign to keep the automatic looms running until the factory hooter announced the end of the work period. Hitherto, and still in the non-automatic sections which openly flanked the automatic looms, the weavers were ready to depart with their looms stopped by the time the hooter went for meal breaks or the end of the day. It was not uncommon for a weaver's looms to be stopped several minutes before the official time.

The management's rationale for this change in the way of working of this section was that such possible losses of one to two minutes in production from a sett of automatic looms was much more serious than from the smaller setts of non-automatic looms. The campaign was met by general hostility and resentment from the automatic operatives, who believed that they already had heavier loads with their multi-loom setts than the non-automatic weavers.

This attitude found expression in action when Laycock spoke to one of the women weavers who had her looms stopped before a meal break some days after the new campaign had been launched. The weaver retaliated by suggesting that the management ought to start on the non-automatic weavers rather than the automatic ones. At the beginning of the next week, with no prior warning, she gave notice of leaving. This came on top of a similar notice – expected in this case – from the other woman weaver, who was to be married in a few days.

This sudden loss of a second of the four experienced automatic weavers shook the confidence of the top management in the 'hooter' campaign. One supervisor said: 'To work to a hooter is such a difference for them when they've never been used to it. It's going to be hard to get people to do it.'

The other operatives in the section saw this incident as a vindication of their attitude, and the obvious hitch in the management's overall plan as justice in the situation. Some of them commented:

> 'They (the management) got all they deserved.'
>
> 'If you push people about enough these days, you must expect them to retaliate.'
>
> 'What the weavers ought to do is to keep them running till the hooter goes, and then just walk off and leave them.'
>
> 'If we are paid to run the looms, stopping and starting them

is part of the work, and it is not right that they should think we would do it in our time.'

'It's not two minutes which are going to make for high efficiencies – it's the general atmosphere and how people are treated.'

The management ceased their campaign, and tried an overlooker's suggestion to wire eight looms from a single switch. This procedure was ultimately adopted for the new looms with the arrangement that the looms ran till the hooter, whereupon the weavers were free to leave and the loom motors were turned off by ancillary operatives detailed for the purpose.

THE NEW LOOMS

The new looms began to arrive on 21 January, evoking considerable interest and excitement at all levels of the organization. The engineers and labourers immediately unpacked them and began installations in the main shed. As soon as they were fixed in position, Irwin and Young, the two original overlookers in the section, left the 48 looms and began assembling and running in the new looms, successively for 16 hours without and with shuttles, and then with a first warp.

The remaining overlookers and the weavers were reshuffled again between the 48 looms.

One of the non-automatic overlookers – a man of 58 – who commenced training in the section in October, returned to work at the end of January after an illness of eight weeks. However, he did not return to the section but took over a sett of looms in the non-automatic department devoted to specialized weaving. This department of 212 looms was at the left-hand end of the main shed in *Figure 1* and was not technically affected by the present innovation. An overlooker from this department now joined Young and Irwin as a trainee for the automatic section.

More looms continued to arrive until, by the middle of February, there were 70. By this time all seven overlookers in the section were working overtime for two hours per day on the assembling work. Delays in the delivery of the British motors for the looms and faults in their design caused still further delays in getting some of the new looms running after they had been assembled.

When the first 16 of the new looms had the running-in warp gated, two of the trainee weavers – formerly responsible non-automatic weavers – moved on to them and the remaining weavers in the section of 48 were were reshuffled. Other changes at this time were the loss at an hour's notice of one of the trainees, and the arrival of two juvenile male trainees from the non-automatic department who went with the two remaining experienced weavers for training in the automatic procedures. The eighth overlooker who, as mentioned earlier, had been delayed since January in the non-automatic department by increasing shift work and absenteeism, also joined the section.

The number of damaged packets from the 48 looms had increased again since the arrival of the new looms. Laycock explained this in terms of loss of the experience of Irwin and Young. The remaining overlookers also felt their loss in terms of recurrent trouble with the looms which had increased considerably. As one overlooker said: 'You fix the looms and a few minutes later, you are back again at it.'

The successive losses through labour turnover of weavers and prospective weavers from the section upset Oakroyd's original plan. Two men – hitherto operators of the drop-wire machine – were encouraged financially to enter the weaving school for a brief introduction prior to joining the section.

By 7 March, 32 of the new looms were weaving the running-in quality – a viscose coat-lining material which was weaving well with a very low factor on the work-loading scheme. Accordingly Oakroyd put them on the bonus scheme. The four weavers who were now on these looms complained to him through their union representative, Monks, that they would drop in wages. After some negotiation an elevated load factor arbitrarily fixed, was used to calculate their bonus earnings and all four did, in fact, improve their wage.

Downham, the experienced automatic overlooker, now had a full sett of 40 of the original looms on one shift and opposite him were two of the trainee overlookers, while the other five overlookers were responsible for the other running looms together with the continuing work of assembly.

The delay in the arrival of the looms, and the character of the nylon warps and wefts in many of the remaining non-automatic looms, had decreased the normal pressure on the preparatory departments except that of sizing. This latter was still fully occupied because the shortness of the nylon warps kept the actual number of units for sizing at a high figure. Natural recruitment in these departments was stopped and normal losses of personnel brought the operatives to the number required.

Another large order for a nylon cloth was obtained at this time by the company. Oakroyd decided to put this quality in the automatic looms following the running-in warp of six packets. The weaving management particularly welcomed this order, since it meant a continuity of conditions which would simplify the remainder of the innovation period.

By the middle of March, 96 new looms had arrived and 56 were weaving. Some of the weavers joined the overlookers on overtime work and helped to unload and unpack the crates of looms. This overtime work helped to boost the wages and the morale in the section, both of which had dropped a little as the weavers took sole charge of setts of looms and as the inexperience of the overlookers began to show itself in increased loom stoppages.

WOMEN WEAVERS

Two women weavers from the specialized non-automatic section joined the section at this time together with another male trainee from the weaving school. The reversal of the management policy on male weavers in the automatic section was caused by the urgency of the situation following the unexpected losses. The two women were asked by Ashby and Laycock to change to the other non-automatic department on shifts, thereby releasing two male weavers. Under this pressure the women agreed to work on shifts – but in the automatic section – since this would mean an increase in wage which the other move did not. The management accepted these terms, and in less than three weeks all three of the newcomers in the section were responsible for setts of the new looms – the two women on bonus conditions.

THE OVERLOOKERS IN DIFFICULTY

With 142 looms in the section, all the trainee overlookers were in charge of sub-setts of looms and their lack of experience was increasingly evident in loss of both production and quality.

To meet this situation Laycock asked Oakroyd to obtain some help for these overlookers. He arranged with the company management that Kaye, previously an overlooker at Debenham Mill and now on the headquarters staff of the company, should be loaned for a month to help out in the section and continue training the overlookers. Kaye had, in fact, run a weekly class of instruction on the manual of the automatic looms for these trainees prior to Christmas. Further help was more difficult to arrange, but through Laycock's influence at Debenham Mill, Harding, the manager there, agreed to exchange for three months one of his experienced overlookers, Knight, for the newest of the trainee overlookers. This increased enormously the overlooking experience at Radbourne and enabled the trainee to continue learning the automatic procedures under the easier and more stable conditions at Debenham.

The nylon quality was weaving well in the new looms and the efficiencies rose rapidly, enabling all the weavers on bonus earnings to improve their wage.

Towards the end of March, the final 16 looms arrived; but motors were not available for these until during May and June, and the last of the 160 automatic looms began weaving on 17 June 1955.

AUTOMATIC WEAVING

At the end of March, 106 looms were weaving, with nine weavers on one shift and ten on the other. All these weavers were now alone on full setts or sub-setts of looms, except for one weaver who, though on bonus conditions, retained another of the weavers on his sett. This weaver – a juvenile – after several months' training in the section in 1954, had taken responsibility for a full sett when the dress goods quality was in the looms at the end of that year. Since the departure of this easily-handled quality, he had experienced more and more difficulty, and his efficiencies dropped while his damage rate rose. In these weeks it was common to see as many as 6 of his 16 looms idle at a time. Finally, in April, Laycock moved both him and another unpromising weaver on the other shift out of the section to jobs in the non-weaving departments of the mill. One of the two former operators of the drop-wire machine was brought from the weaving school, to make nine weavers on each shift.

A new sizing machine was installed between the new beamer and the other sizing machines. This was particularly necessary now that so many of the automatic looms were weaving nylon qualities. The narrowness of the cloth compared with the wide automatic looms brought new problems for the overlookers. However, the yarn's breakage rate was low and the spools lasted a long time, so that a weaver had a work-load of only 37 on 16 looms as calculated by the work-study methods. This meant that the possible bonus earnings would have been lowered with even the highest efficiencies. The weavers again complained against this threat to their wages, and the management agreed to an arbitrarily fixed load of 80 for setts of 16 looms.

WEAVING NYLON CLOTH – A STABLE STATE

By mid April, all overtime work had ceased, and production and quality from the new automatic section were very satisfactory. The managers nevertheless qualified their satisfaction in terms of the easy quality in the looms and the external help in the overlooking. Kaye returned to the company management at the beginning of May, and the eight overlookers were alone on setts of looms, although most of these were below the full size of 40 looms.

After another three weeks of these stable conditions with improving performance, the management decided to introduce two other ways of working related to automatic production. First, the size of the weavers' setts was to be increased to 24 – an approach to the size of 32–40 which was calculated from the work-study measurements on the nylon quality. Second, the looms in the section were to be run throughout the meal breaks. At other times in the history of the mill, the management had tried to run looms through the meal breaks, but the subsequent rise in the damage rate had nullified the gain in production. Now, however, the management based this attempt on the automatic rationale of increased continuity, as well as the practical suitability of the nylon quality for such an experiment.

SETTS OF 24 LOOMS AND CONTINUOUS PRODUCTION

The first setts of 24 looms appeared in the last week of May; and when all the looms were running, there were six such setts and one sett of 16 looms. This decrease in the number of setts made one full weaver on each shift surplus.

The management had expected little resistance to the increased setts from the weavers in general, because of the generous conditions that they had enjoyed over the last two months and the further financial incentives that were to go with the change. Moreover, apart from some grumbling, the change was easily accepted. However, in the redeployment of the redundant weavers the management were in a difficult position. Laycock had given most of the new automatic weavers verbal assurance that their change to the section would be a permanent one. Laycock asked two weavers in the section to take 20 non-automatic looms weaving the same nylon quality. One was the last non-automatic weaver to join the section and he agreed to return to his old section. The other – one of those specially recruited by the company in 1954 for the automatic section – refused to change unless compensated for any drop in wage he might experience. He maintained that this could arise in two ways. First, because all his experience, apart from a short time in the weaving school, was of automatic looms. Second, although his work-load on the non-automatic sett would be larger than on an automatic sett, the wage systems in the two sections were now so different that it would be impossible for him to earn as much. His union representative supported the weaver in his stand and, after some negotiation, special compensatory terms were offered to both the redundant weavers, and these were accepted.

For the setts of 24 looms, a new sort factor for the nylon quality was introduced which gave a work-load of 84 against the 'arbitrary' load of 80 for 16 looms. This offered only slightly improved bonuses to the weavers if they maintained their sett efficiency through the change. However, at the same time, the target wage was raised by 5½ per cent. For some weeks while the setts were changed, two load factors and two payment schemes existed in the section for the same quality of cloth.

For the running through the meal breaks, the management again planned to offer a financial incentive to both the weavers and the overlookers. The meal breaks were to be staggered over an hour so that a single weaver and an overlooker were in attendance, on 48 and 80 looms respectively, for each half hour. In practice, by moving the trainee weavers in the section from sett to sett for this hour, no weaver was alone on 48 looms until July. The extra picks were simply added to the weavers' totals and the efficiency calculated on their working time of 37½ hours instead of the new theoretical time for the loom of 40 hours. This meant an increment for the weavers of approximately 70 per cent of their normal rate. The overlookers were offered a fixed increment equivalent to 43 per cent of their hourly rate while they had charge of the 80 looms. The weavers accepted these terms, but the overlookers rejected them on the grounds that they had a double load when they were responsible for the 80 looms. The management replied that this was not so in terms of the actual work they could do in this time. A compromise was agreed upon whereby the overlookers' increment was 70 per cent of their normal rate.

Both changes proved to be very successful. The actual efficiencies of the setts were maintained at their high level, while the damage rate did not rise appreciably. This increased production put heavy pressure on the preparatory departments. In particular, the spooling department was unable to supply enough nylon spools, and instead of the weavers loading the magazines from a board of spools on each loom, they were often working as many as 8 looms from a single board. To keep their looms running, the weavers were even fetching spools from the spoolers, although this was not allowed by the management.

In mid June, Knight returned to Debenham Mill and the overlooker who changed with him returned to the section on a full sett of looms.

THREE SHIFTS OR SATURDAY WORK

Early in July, Oakroyd was told by the company production manager that he would have to increase his output of the nylon quality still further, so that looms at Debenham Mill could be freed for other orders. Two possible ways of achieving increased production were considered. The first was to put one or two overlookers' setts of the automatic looms on to three-shift working, which would involve up to ten male operatives on a hitherto unknown night shift from 10 p.m. to 6 a.m. The second way was to work all the personnel in the section, together with numerous others from other departments, on alternate Saturdays.

Laycock sounded out several of the key operatives about these two possibilities and met very strong opposition from the overlookers about the three shifts. They claimed it was impossible, because their national union had recently taken a stand opposing such continuous work. The management did not strongly press the issue, since the national weaving unions were in fact at that time discussing the whole question of three shifts in the industry. Accordingly, despite its economic disadvantage, the management decided to adopt the second course. The unions accepted this proposal after securing agreement to their request that work should cease at noon on Saturday. The first overtime was on 10 July, just three weeks before the mill closed for two weeks – the annual holiday. Saturday-morning work continued afterwards until the end of October, when a five-day week of 37½ hours was resumed with the looms running through the meal breaks.

During July, the number of operatives of all types in the automatic section was sixteen or seventeen, which gives a figure of 9 or 10 looms per operative. The corresponding figures for the other weaving sections in the company are given in *Table 3*.

TABLE 3 TOTAL NUMBER OF LOOMS PER OPERATIVE

Weaving Department	*Looms per Operative*
Radbourne Mill	
Specialized non-automatic	3·1
Old automatic	5·1
Non-specialized non-automatic	8·3[1]
New automatic	9·5
Debenham Mill	
Automatic	8·5

[1] At that time weaving a specially smooth-running nylon sort.

The downtime of looms in the section continued to fluctuate widely, with an average of 19 hours, which was three times that at Debenham Mill.

THE END OF THE WEAVING SCHOOL

At the end of August, the group of non-automatic looms in the corner of the main shed that had formed the weaving school reverted to production weaving. The two remaining trainees moved on to some looms in the redrawing shed adjoining the maintenance shop. A non-automatic weaver from the specialist department supervised them and her production looms were taken over by Miss Harrison, who had hitherto been in charge of the corner weaving school.

NEW QUALITIES IN THE SECTION

From the middle of October, the nylon quality began to fell out from the automatic looms. Until this time, the total personnel in the factory had been slowly decreasing, as described earlier, but now recruitment began again.

Without the nylon quality the quasi-automatic operation in the non-automatic section could not be maintained – the size of the weavers' setts would be markedly reduced – and a shortage of weavers became apparent. A lining quality, which had rather higher work factors than the nylon quality, was placed in most of the automatic looms. Despite work-loads of over 100, the managers maintained the setts at 24 looms. The efficiencies dropped somewhat, but the weavers' wages remained at the same level because of the higher loads. One of the promising juvenile weavers in the section left in October, and his place was taken by yet another woman weaver from the specialist non-automatic department.

In November, the overlooker who had earlier exchanged with Knight at Debenham Mill, was permanently transferred to fill a vacancy which had arisen at that mill. His place was taken by an overlooker from the old automatic department, who had had some experience of the new automatic looms elsewhere. A reshuffle of the overlookers in the section with respect to their shift pairings was made to coincide with this change. This reshuffle was instigated by the overlookers themselves and worked out jointly with the managers.

A few dress goods qualities with very high work factors were beginning to appear in the section. Some attempt was made to distribute these throughout the section so that the setts of 24 looms could remain, but eventually, early in 1956, setts of 12 looms were set up for these qualities.

PREPARATORY DEPARTMENTS UNDER STRESS

These changes in quality were now being felt acutely in the preparatory departments. Their depleted staffs could not be increased quickly enough to avoid serious shortages of both weft and warp during the last two months of 1955. During the third week of November, on a number of occasions looms in some of the automatic sets stood idle because of the weft shortage. The following comments are typical of this period:

> 'I've never seen things worse than they are now. It's absolute chaos, and we [the weavers] are chasing weft all over the place.'
>
> 'I go to the spooling department myself to get weft to keep the looms running.'

As the looms felled out, there were no further warps ready and the downtime of the looms increased enormously. The last week of November saw three setts of 24 looms with 9 looms empty. At least one loom stood idle for the whole of the last week in November. Things were even worse in other weaving departments, for preference had been given to the needs of the new automatic section. More than 100 looms – half the total – stood idle at one time in the specialist non-automatic department. Just before Christmas most of these quality changes were complete and the efficiencies in the section began to rise again. *Figures 6* and *7* show changes in the efficiencies, quality, and quantity of production in the section over the innovation period, and *Figure 8* shows the downtime of the looms and the number of warps changed per month.

Fig. 6: *Efficiency and Quality of Production in the Automatic Section*

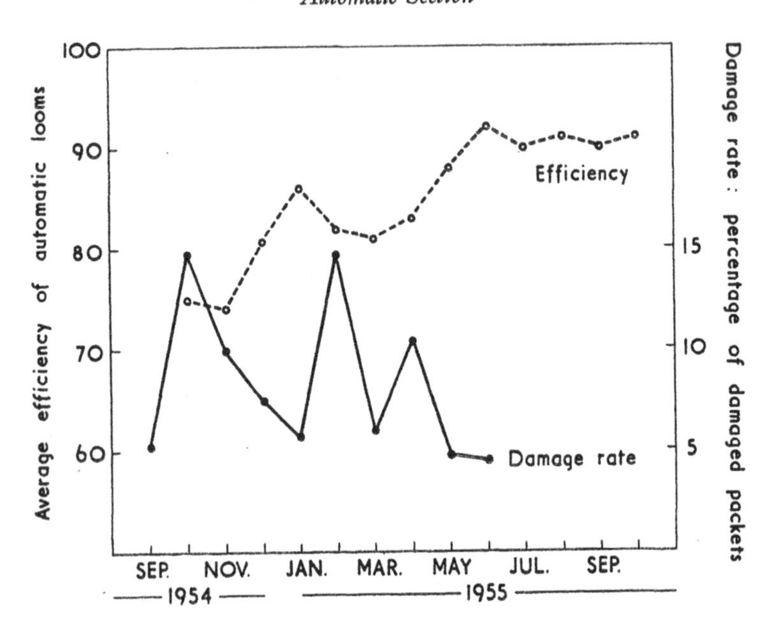

Fig. 7: *Total Production of Cloth in the Automatic Section*

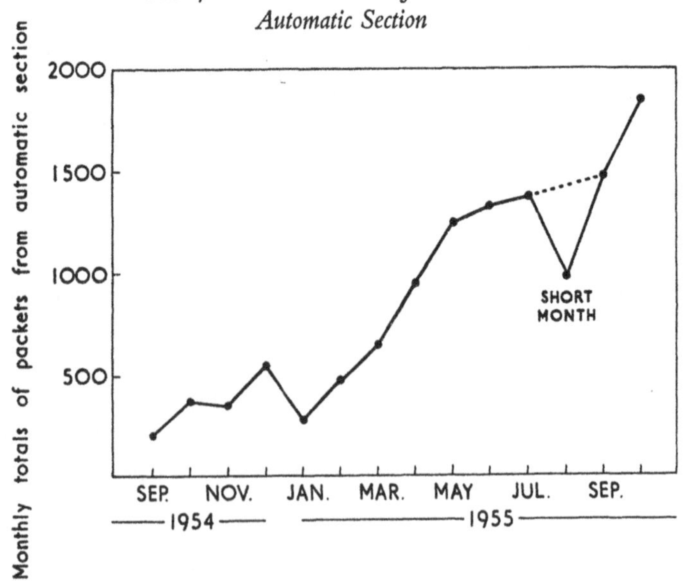

FIG. 8: *Turnover of Warps and Loom Downtime in the Automatic Section*

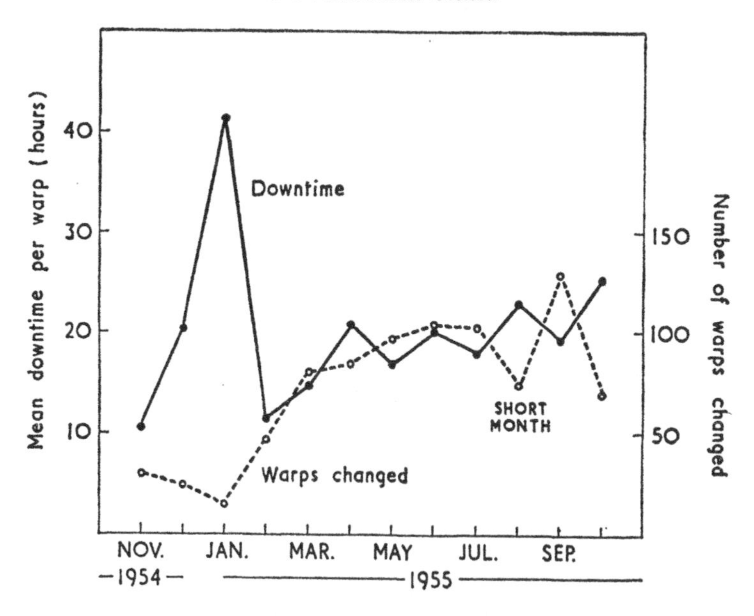

FURTHER CHANGES IN THE MANAGEMENT STRUCTURE

Two further changes in the management structure occurred at the end of the year. The foreman of the non-automatic weaving department, which had lost looms to make room for the automatic section, was to spend some time as a quality supervisor in the automatic section. This, however, was not a popular move and lasted only a short time. Watson, the foreman of the sizing department, returned to the position of weaving instructor that he had held for many years prior to 1953. He took over the temporary arrangement for training weavers in the redrawing shed and began to develop it as a comprehensive school in which a number of recruits were enrolled. Oakroyd did not replace him in the mill but extended the responsibility of Black, the warping foreman, to cover both the sizing and the warping departments.

Throughout 1955 a variety of automatic spooling machines had

FIG. 9: *Reorganized Layout of Radbourne Mill by early 1956*

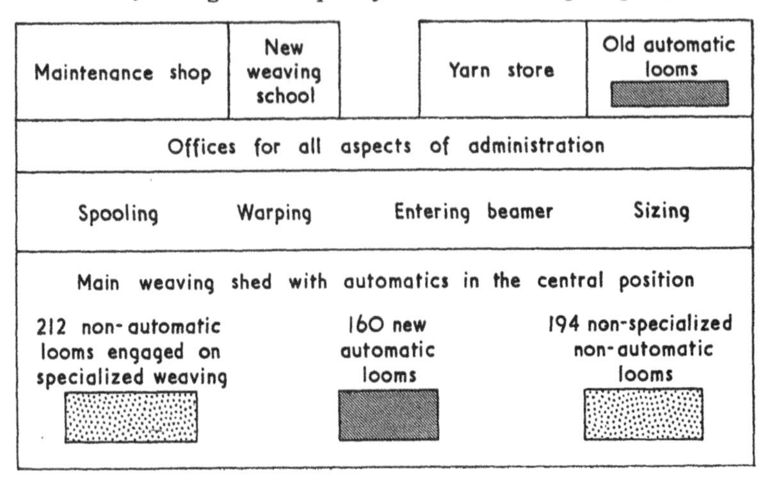

FIG. 10: *Formal Management Organization of Radbourne Mill in early 1956 after the Innovation*

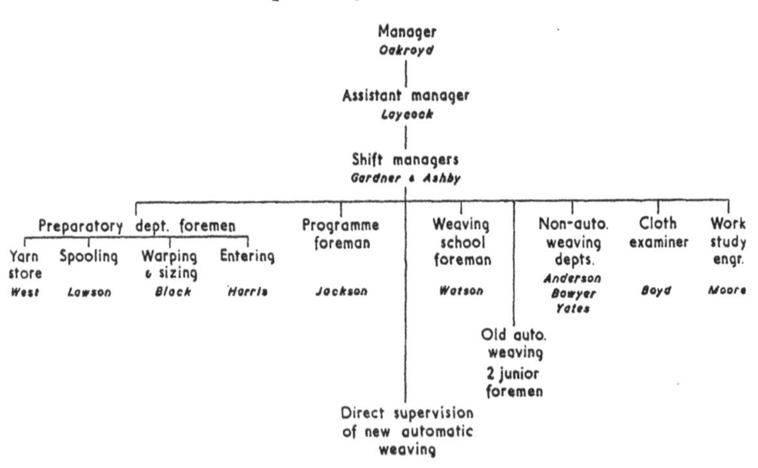

been thoroughly tested and, in fact, they contributed to the output of the department. However, no general re-equipment occurred in this department.

This study effectively ended at the beginning of 1956. By this time, all the new looms had been in operation for at least six months.

The geography of the mill had undergone several changes and now appeared as in *Figure 9*.

The formal structure of the management group had also changed since the beginning of our study. It now appeared as in *Figure 10*.

SUMMING UP

In this chapter we have described the events which actually took place in connection with the innovation, and for ease of reference these events are presented in tabular form in *Table 4*.

We were able to distinguish several stages in the period which we studied. First, there was the preparatory stage before the looms actually arrived when the older automatic looms already in the mill were moved into their new situation. This was probably a valuable period, since it enabled some social changes which had been made to be tried out in the new social and technical setting, and we have described some of the tensions that arose during this period.

Second, there was the stage covered by the arrival and installation of the new automatic looms. This was accompanied by the major social change of appointing two shift managers. During this period everyone was concerned to get the new looms working efficiently and effectively as soon as possible, and little obstruction of this immediate goal seems to have taken place.

Lastly, there was the stage when the looms were doing adequate production work, and we were able to see the people in the mill coping with the demands of the new technology under the stable conditions of the nylon weaving programme, followed by the disruptive effect of changing over to weaving other fabrics. During this period, too, social changes of an adjustive character were taking place in relation to some of the needs of the new looms.

TABLE 4 CHRONOLOGY OF THE SIGNIFICANT CHANGES IN THE AUTOMATIC SECTION DURING THE INNOVATION PERIOD

1	Sept. 1954	48 automatic looms in original position (see *Figure 1*).
2	Oct. 1954	These looms moved to main shed (fixed wage).
3	24 Oct. 1954	No humidification in section.
4	21 Nov. 1954	New humidification plant begins operation.
5	8 Dec. 1954	Bonus wage conditions again.
6	21 Jan. 1955	New looms begin to arrive (two most experienced overlookers leave production looms).
7	Apr. 1955	Nylon quality begins to appear in section.
8	1 June 1955	Weavers' setts increased to 24 looms and looms run through meal breaks.
9	17 June 1955	All 160 automatic looms now running.
10	14 July 1955	Saturday-morning work begins.
11	20 Oct. 1955	Nylon quality begins to fell out from the looms and new qualities with higher sort factors appear.

FIVE

Management Under Change

We have already noted that Radbourne Mill had a much longer history of weaving than had Debenham Mill. This fact was important in many spheres and not least in that of management. By management in this context we mean the managerial group exercising supervision and control. This group is shown in *Figure 5* on p. 42.

In 1873, there were slightly more than 1,300 people employed at Radbourne; during our study, the figure was around 400 people. Thus, though the physical size of the mill had, if anything, increased, its population had diminished drastically. Even so, at the time of our study, though the management group was considerably smaller and less complex in function than it had been in the late nineteenth century, it was nevertheless more complex than its contemporary equivalent at Debenham.

As we shall hope to show later, the historical aspects of supervision in the mill played an important part in determining some of the tensions that subsequently arose out of the innovation we witnessed, and we, as observers, were made much more aware of the recent past at Radbourne Mill than we had been at Debenham.

STRUCTURE AND FUNCTION OF THE RADBOURNE MANAGEMENT GROUP

The formal structure is indicated in *Figure 5*, p. 42. This shows the management structure which existed just prior to the innovation period. Certain changes were made during the period of our study and these will be referred to later. In September 1954, when the study began, the total personnel numbered 409, of whom 391 were operatives and clerical and administrative staff.

Taken as a group, the foremen and managers were not as homogeneous as the Debenham group. There, the whole group was recruited from local men, most of whom had spent virtually all their working lives in the company, and in the same part of the country. They had not necessarily been at the same mill all the time, but 'ecologically' they were very similar. At Radbourne this was not the case. There, of the management group which just preceded the innovation, six were from the North of England, and five of these had come to the Radbourne area since the war. Both Oakroyd and Laycock, the manager and assistant manager, came from the North. Most of the group had, however, spent a considerable portion of their working lives with the present company.

Structurally, it is more meaningful to present the organization of the mill in a way other than that of the formal hierarchy, and this is done in *Figure 11*. This way of presenting the organization blurs the status differences between the levels of management, but brings out the functional relationships of the departments.

This figure can be compared with *Figure 2*, p. 28, which represents the Debenham structure in the same way. It is immediately apparent that there is considerably more differentiation of function at Radbourne. This is particularly noticeable in the preparatory sphere, where there are five supervisors each with a different department, compared with one large department and one supervisor at Debenham.

This figure also shows that, before the innovation, there were three fairly distinct status levels within the management group – manager, assistant manager, and supervisors. In addition, the section of automatic looms which was already in existence in the mill, had a senior supervisor with two 'junior' supervisors under him – one on each shift. These junior shift supervisors still had some production tasks – each was the responsible overlooker for a small number of looms.

FIG. 11: *General Management Organization of Radbourne Mill*

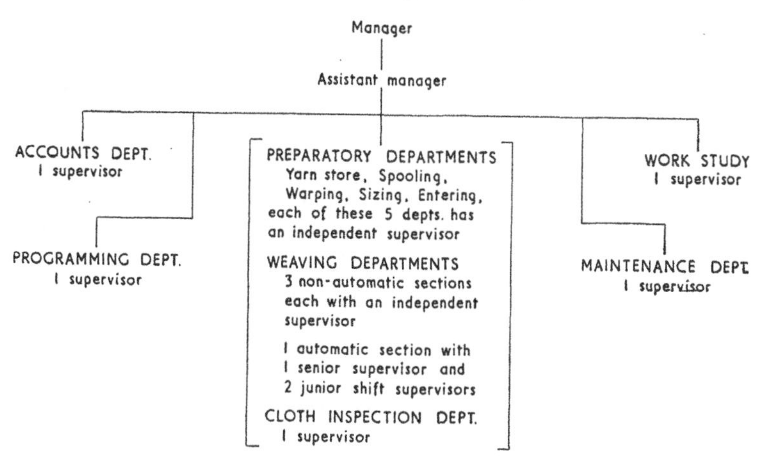

These organizational differentiations were, on the whole, in conformity with the physical barriers existing in the mill. These tended to separate one preparatory department or weaving section from another, except in the area of the 'big shed', as it was called (see *Figure 4*, p. 41). Here, before the innovation, there were two large sections of looms without a physical barrier between them. In addition, the sizing and entering departments were separated by space rather than by a barrier of any sort. The manager and assistant manager had separate offices.

Before the innovation, then, differentiation of function was at a maximum, with service and production supervisors clearly separated from each other. Historically, this differentiation had been emphasized in other ways. For example, warping had always been regarded as highly skilled work, and was formerly one of the major repositories of company secrets of novel types of cloth. Hence the warping department was particularly isolated and differentiated from other departments and from the outside world. Although these conditions no longer strictly held, the operatives still felt and talked in terms of the distinctiveness of each department and cherished the skills of the best workers as contributing to this.

INNOVATION

What did the innovation portend for the management group? We have outlined the major characteristics of automatic looms, and some of the qualitative answers to this question will also be given in other chapters dealing with operative skills. But is it possible to attempt any quantitative assessment of the changing conditions? One possible approach is to compare the productive capacity of the mill under two sets of conditions. A possible measure here, which is independent of the type of cloth woven (and thus of difference in yarn thickness, pattern complexity, and so on) is the index of potential loomwork. This is achieved by calculating the total number of possible 'picks' to be obtained from all the looms in the mill. (The reader may recall that a 'pick' is one crosswise movement of the shuttle across the warp threads in the loom.)

Table 5 shows this comparison just prior to the innovation, and also at the end of our study.

TABLE 5 PRODUCTION POTENTIAL, RADBOURNE MILL

Date	*Total No. of Looms*	*Total Possible Picks per Week ('000s)*	*% of Total Picks Woven on Auto. Looms*
October 1954	723	303,732	30
September 1956	688	327,447	53

Assessing these raw figures, we see that, in the two years which passed from the beginning of the innovation, while the actual loom strength dropped by 5 per cent, the total picks possible increased by 8 per cent. In addition, the balance between automatic looms of all kinds and non-automatic looms of all kinds had swung towards the automatic looms in terms of productive capacity, although in terms of actual numbers of looms there were still more of the non-automatic type.

These changes in weaving capacity and in the distribution of capacity between automatic and non-automatic looms do not necessarily result in comparable changes in the work of preparation departments. The primary factors governing increases and decreases of work in these departments are:

(*a*) Thickness of yarn in the spooling department. The thicker the yarn, the greater amount of work, since the length of thread on a pirn is determined primarily by yarn thickness.
(*b*) Length of warp in the warp preparation department. The major part of the work here is concerned with setting up the warp, so that two warps of 100 yards length take a great deal longer time to process than one warp of 200 yards length.

Thus, these two additional factors – thickness of yarn and length of warp – could well conceal actual increases in work which must arise through an increase in weaving capacity of 8 per cent. The increase in demand arising from greater weaving capacity might well be overlaid by these other factors, as well as by the fact that automatic looms are only one section of the total number of looms in the mill, so that no easy comparison is possible. This must not be taken as meaning 'no change', but only a change very difficult to quantify.

MANAGEMENT CHANGES FOR THE INNOVATION

Two important steps were taken by the company to help the management group to deal with the problems that the innovation was expected to bring. In July, Laycock was appointed to Radbourne as assistant manager. He succeeded a much younger man who had only recently completed his training as a graduate trainee. This man had been not 'assistant manager' but 'assistant to the manager' which implied a considerable difference in responsibility. The assistant to the manager was engaged on particular aspects of the manager's task that were specifically delegated to him, whereas the assistant manager carried full responsibility for some aspects of the work of the mill.

Laycock, the new assistant manager, had been senior weaving supervisor at Debenham Mill during the period when that mill was changing over to automatic looms. Prior to that installation he had spent some time in the U.S.A. with the manufacturers of the looms concerned. For several years following his Debenham experience and before his appointment to Radbourne, he had been working at the headquarters of the mills as a specialist in quality control. This man, therefore, brought to Radbourne considerable knowledge and experience of automatic weaving. However, although this experience had been a telling factor in his selection, the carrying through of the innovation was in fact only a part of his overall responsibility as assistant manager for all parts of the mill.

The second important step was taken towards the end of 1954. When the first new looms were just about to arrive, Gardner, the automatic loom shed foreman, and Ashby, a non-automatic loom foreman, were each given the post of shift manager. Up to this time they had shared responsibility for the preparations in progress on the new automatic section. Their new title was 'shift manager' and they were formally in charge of the whole mill during the period of shift work before and after the normal day-work hours. Both men were long-service employees of the company who had spent many years working at Radbourne, and Gardner, of course, had had experience of the already existing section of automatic looms. The creation of this new post of shift manager meant that one man was promoted to non-automatic foreman, and the two junior foremen in the old automatic shed were promoted to full foremen, relinquishing the small sett of looms which they had previously run jointly as overlookers. At this stage, no reorganization was carried out in the non-weaving departments.

Thus the mill entered the main innovation period with two

additional managerial posts (although in effect this amounted to one post, since only one shift manager operated at any one time) and an assistant manager, with clear executive responsibility, appointed to give overall guidance on the installation and integration of the new looms. The other appointments arising out of the appointment of the two shift managers were made within the existing structure.

THE IMPACT OF THE INNOVATION

The reader is referred to Chapter Four for more details concerning the actual day-by-day events of the innovation. Here we wish to show how this innovation was perceived, in particular by the supervisors, during the innovation period and afterwards. Here, too, we shall be more concerned with supervisors other than the shift managers, since they were – in some degree – more able to form realistic expectations of what the new looms would mean.

As far as the preparatory departments of the mill were concerned, there had already been some introduction of more modern machinery. In the spooling department, the sizing department, and the warping department, new machinery – which by no means affected the bulk of work in these departments – had recently been installed, partly in order to deal with the two major requirements of the new looms: an increase in quantity and in quality of materials produced.

Initially, the effect of the new looms on the preparatory departments was different from one to another. Lawson, the weft department supervisor, thought that the changes for him were considerable both in relation to quantity and quality. He said: 'The S.6 pirn requires more supervision to ensure that the standard is good, and bad spooling has to be put right quickly.'

Technically it could have been predicted that the first impact of the looms would be felt by this department, since one essential 'automatic' part of the looms is the automatic replacement of each pirn or spool as it becomes empty. If the pirn is not exactly correct then the loom stops and since a change occurs several times an hour, bad pirning would very soon be noticed.

This supervisor had so organized his department that the people doing shift work produced the pirns for the looms running on shift work – which included the new automatic looms.

In the warping and entering departments the supervisors noticed no immediate change in either quality or quantity requirements of their work. 'The warps for the new looms have to be entered in the same way as every other warp and in fact they are not as complicated as some other warps.'

In the sizing department, however, there was definite evidence of increased pressure for quantity and quality of work. As we have mentioned above, the sizing department did have a new sizing machine installed, but there were considerable teething troubles with this machine, which counteracted some of its effectiveness. And, in addition, there was real pressure on this department from the managers of the mill. One of the managers said: 'We are putting more pressure on, and I've been pushing the sizing department . . .'

During the early days of the innovation one or other of the managers could often be seen in person investigating difficulties of production in the sizing department.

As the innovation progressed during 1955–56 the two supervisors, who had initially noticed little change, began to experience the effects of increased quantity of work and, more particularly, of cycles of pressure, when several looms felled out together and required new warps to be put in position. As one of the supervisors said: 'Pressure has increased a lot. No sooner do we get a stock of warps on the floor, than they are taken up by a heavy fell' (i.e. a lot of looms requiring new warps at the same time).

During these periods of heavy activity, another factor of importance emerged within the management group. It will be recalled that Radbourne is a mixed weaving mill in which both automatic and non-automatic looms were working, but when there was competition for warp preparation machinery, generally the automatic loom warps were given priority. One supervisor perceived this new situation in this way: 'In terms of quantity, the management don't like the new looms to be waiting, because they're new and they're very costly to run. But we have no *instructions* about priority for the looms. It's just that I know how the management feel about it.'

As might be expected, the supervisors of the older loom sections were not entirely able to agree with this policy despite its economic soundness. One of them said: 'There's definitely a preference for the new automatics for both warps and spools, so that if they get short we're left high and dry. We're always waiting for something. We're lost in the queue.'

Can we advance any explanation for the fact that the spooling supervisor thought that the new looms would bring different demands, whereas the warping supervisors felt the differences only over a period of time? Two explanations can, in fact, be offered. First, the reader has probably realized that the relationship between spooling and weaving, and warps and weaving is different. Schematically, it can be illustrated as in *Figure 12*.

This illustrates how the production of spools was much nearer to the weaving function in every sense than was the initial production of warps, and Lawson's earlier perception of the problems to be expected must have been derived from this fact. Second, a factor of considerable importance was that inadequacies in the quality or quantity of spools was felt immediately by the weavers and overlookers, and this information could be quickly transferred to the place in which the total preparation process took place, i.e. the spooling department.

At a fairly late stage of our inquiry a change in policy within the mill enabled the warp preparation line also to become more coherent. This move was dependent on a whole set of factors which we must briefly outline. Shortly before the period of our study in 1955, the weaving training school within the mill had been discontinued as a unit, and weavers in training had gained experience on a small number of looms in the main weaving

FIG. 12: *Production of Warp and Weft at Radbourne Mill*

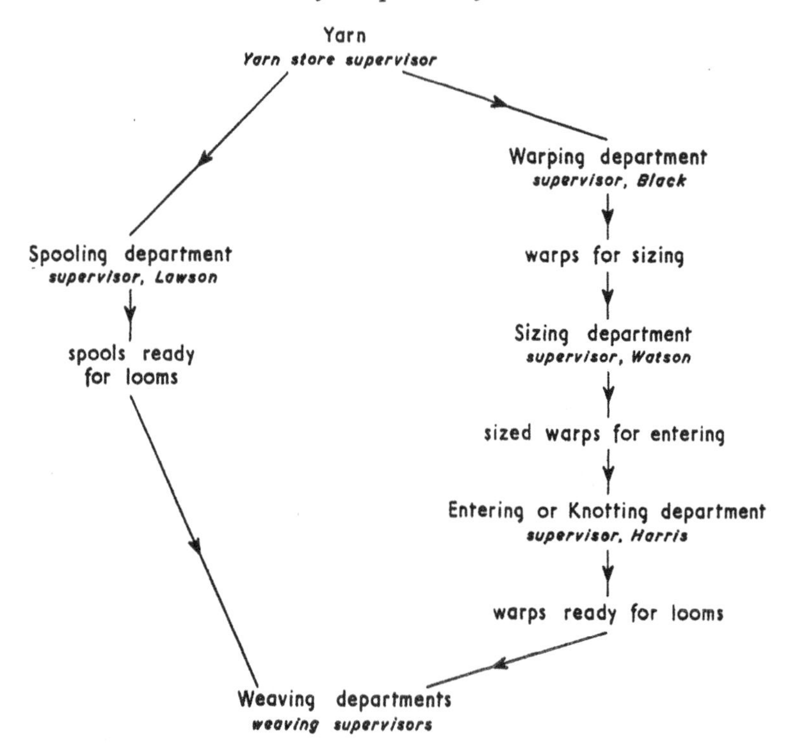

department. During the period of the study, the shortage of adequately trained weavers became more and more pressing until finally it was decided to re-open once more the old weaving school as a separate unit. The person who had previously been the weaving instructor was Watson, now the sizing supervisor. Early in 1956 – about a year after the start of the innovation – Watson was asked to start up the full-time weaving school again.

This left the sizing supervisor's post empty, and as a consequence the opportunity arose to adopt a rather different policy with respect to the supervision of this department. Black, the warping supervisor, was asked if he would take control of both warping and sizing departments, and this he did. It is possible to see this new appointment as a direct attempt to integrate the warp production more closely in order to control more effectively the supply of warps to the weaving departments.

Black himself perceived this reorganization as a definite improvement, and made this comment: 'It's easier to run the two departments together as far as planning goes. When I'm asked to warp something urgently, I can make sure there's a machine ready for it in the sizing department.'

Certain unforseen circumstances arose out of this change. Black inevitably spent a considerable part of his time in the sizing department learning the new techniques in detail and tackling the operating problems. This meant that the warping department was largely left to 'run itself'. Unfortunately, it did not do so, and the workers in the warping department felt the lack of an immediate supervisor for communication purposes. This situation persisted for several months in which the warping workers used other people as supervisor-substitutes. These substitutes included the programming supervisor and a senior programming clerk, and after some months of these conditions the senior clerk was appointed as junior supervisor in the warping department to fill the gap which had emerged as a result of Black's new role.

THE PROGRAMMING FOREMAN – NO. 3 OR A MINOR ROLE?

The difficulties arising from the integration of the warp production programme lead us now to consider a role with which we have not so far dealt, that of the programming supervisor, Jackson,

At Radbourne the short-term programming and planning function was the overall responsibility of Laycock, the assistant manager. Laycock's office was in close proximity to a larger office known as the 'order office', which was under the supervision of Jackson. Jackson acted within a programme which was prepared weekly by himself and Laycock. This programme was known as the *fell forecast* and showed the production position in relation to (a) supplies of yarn, (b) progress of orders, (c) expected 'fell', (i.e. end of the warp on a loom) (d) new assignment for those looms which were expected to fell. This programme was arranged on the basis of the present state of weaving in the mill, and on the changes which would have to be made to fulfil the new orders for woven cloth received from the central administration for all the mills. It was one of the main tasks of the assistant manager to ensure that as much of the programme was carried through from week to week as proved possible in the light of the changing conditions of personnel and materials.

Laycock considered Jackson a vital link in ensuring that this process was carried through and in fact referred to him as the third man in the mill. This was in many ways an exaggeration of his status in the mill, but, from our observation, he had an important part to play in coordinating the activities of the preparatory departments. From previous experience Laycock well knew the importance of coordination of activity for weaving with automatic looms and it was this experience which led him to place such great emphasis on Jackson's activities. Nevertheless, such coordination was in conflict with the preparatory supervisor's traditional autonomy of action. In the introduction to this chapter we noted the probable importance of historical factors in relation to the innovation, and autonomy of supervision was one such factor with strong historical roots that tended to militate against efficient use of the automatic looms.

Laycock was well aware of these current attitudes and worked in several different directions to decrease the influences isolating one department from another. Concerning this very point, he at one time commented: 'They don't mind you trying things and they don't resist them. They just don't embrace them and carry them on themselves.'

The reality of these coordination difficulties is sharpened by comparing the data kept by both Debenham and Radbourne Mills on the loom downtime occurring as a result of changing warps. Radbourne's data are shown in graphic form in *Figure 8*, p. 63, and the difficulty is obvious on inspection. In addition, it is interesting to consider the downtime for the older existing automatic section at Radbourne and the comparative downtime figures emerge as follows:

TABLE 6 MEAN DOWNTIME OF LOOMS PER WARP CHANGED: JULY–OCTOBER 1955

Weaving Section	*Mean Downtime (hours)*
Radbourne Mill	
New automatic looms	21·5
Old automatic looms	21·25
Debenham Mill	
Automatic looms	5·75

An incalculable part of the times for Radbourne is of course accounted for by the much greater complexity of the weaving function at Radbourne as compared with Debenham, but some of the difference between the mills is due both to difficulties over supplying the warp and to further difficulties over fixing the warp in position. The situation as regards weft shortage during the same period was quite different, and was so small that it contributed imperceptibly to the downtime figures.

The figures quoted for Radbourne must also be considered against the historical backcloth of the mill. In discussing the length of time taken to refit a loom with a new warp, several of the weaving supervisors remarked that in the 'old days' it was nothing remarkable if looms were empty for several days on end. Indeed, if a particular weaver was absent through sickness or other reasons, the looms in question would quite probably be covered up and inactive until the weaver returned to work on them again. Against these historical facts, the 22 loom hours shown in *Table 6* represent only three shifts or one and a half full working days, and only one-sixteenth of the weavers' total loom hours – comparatively a fairly rapid turn-round of the loom. It is only unfortunate that we do not have the data to compare the equivalent figures for, say, five and ten years previously.

With this type of model as a representative of the 'good' old days, there was obviously a very strong discrepancy between the automatic and the non-automatic work cultures, and the data on the Radbourne downtime illuminate this discrepancy. Finally, another most striking thing about these data is that we calculated the figures from mill records ourselves. No inter-mill, or intra-mill data of this type existed so that a comparison on this basis was obviously not felt to be necessary.

SHIFT MANAGERS AND THE PREPARATORY DEPARTMENTS

We have so far dealt with the preparatory supervisors, but an area of considerable importance is the pattern of relationship which developed between the shift managers and the preparatory supervisors. One of the shift managers, just after he was appointed, showed some recognition of this – at least verbally – and commented on the situation: 'I've been more connected in the past with the weaving shed than I should have been. In fact, I didn't take the opportunities which existed to learn about other departments. I always regarded preparatory people as a means to an end. This was wrong and I've got to alter now.'

This statement is both positive in feeling and reflects an accurate understanding of the new ways of behaving which would be required by a change in role. However, parallel changes in relation to the new role of shift manager were not observed in the preparatory supervisors, whose roles – explicitly at least – remained much the same. Indeed, there was evidence that they extended their supervision to shift hours when the shift managers had control of the whole mill. For it was common practice in these departments for the supervisor to ensure that both information and material were available for the whole of the shift period. Only in cases of extreme emergency was there any call upon the shift manager to act in a managerial capacity. If we can assume that theoretically the position of shift manager during shift hours was analogous to that of assistant manager during day hours, then the behaviour of the two was markedly different, since the authority of the assistant manager over the preparatory departments was both practised and accepted as and when the need for intervention arose.

It is not easy, however, to achieve the sort of change which is implied by the shift manager's statement. We have already discussed the way in which the preparatory supervisors acted as 'absent supervisors' by leaving everything prepared before they left, and this can in part be interpreted as a refusal to accept and perceive the new role of shift manager. An incident which occurred during the shift manager's hours points up this non-acceptance of the new managerial role, at least during the period of our study. One particular operative was detailed to do certain tasks in a preparatory department other than his own, during shift hours. Inevitably, there came a time when this particular worker had to decide which department had priority in case of conflict. He referred the situation to the management for some policy to be laid down. It might have been supposed that this was the province of the shift manager, since the problem was one which occurred during the shift hours. However, the priorities were, in fact, finally determined by the assistant manager and not the shift manager.

In addition to this observed behaviour, the operating responsibilities of the shift managers also militated against their establishing an overall managerial role. This was because, throughout the period of our study, the shift manager was the first-line supervisor for the new automatic section of looms. His role in relation to the workers on these looms was precisely similar to the role of the weaving supervisors on the other sections of looms. But as we shall show more particularly in Chapter Nine, the roles of 'automatic loom supervisor' and 'shift manager' of necessity often clashed, and at least in the case of the automatic loom operatives, the clash resulted in decreased time being available to attend to the problems of that particular section.[1]

TOP MANAGEMENT

The two managers, of course, adopted different points of view towards the shift managers, and from these upper levels their new position in the management structure was recognized. For example, as we shall consider in greater detail later in this chapter, there was one formal management meeting at which both shift managers were present, and this was a weekly meeting of themselves with Laycock and Oakroyd, the manager. No other supervisor was present. Then, again, during the latter part of our study, a daily meeting was started between some of the preparatory supervisors, Laycock, the shift manager on duty, and Oakroyd. At this meeting, as at the former meeting, no other weaving supervisor was present. Thus, by these new groupings the two shift managers were integrated to some extent with the manager and the assistant manager as people whose roles did not confine them to a single department but who had factory-wide responsibilities.

RELATION BETWEEN THE FORMAL AND INFORMAL STRUCTURE OF THE MANAGEMENT GROUP

So far we have been largely concerned with the effect upon individual managers and supervisors of the introduction of the automatic looms and the changes which were made in the formal structure of the group. It is also important to consider the *group as a whole*, and the way in which it actually functioned in the changing context. These considerations are often relatively unrecognized – at least explicitly – in a social system like Radbourne

[1] These changes in role correspond to the lessening of immediate supervision of their operatives and the increase in interdepartmental concern which characterized the supervisors in the study by Emery and Marek (1962) of automation in a power plant.

Mill, but this does not mean that they are unimportant, nor that they are implicitly unrecognized.

The social conditions inevitably invited differentiation within the management group, since it was composed of three formally recognized status levels, later to be increased to four when the shift managers were appointed. In total, this group contained 19 individuals; 4 in managerial positions, 5 as weaving supervisors, 6 as preparatory supervisors, and 4 as non-production specialist supervisors.

In *Figure 10*, p. 64, we gave a representation of the functional aspects of the formal organization. These are, of course, based on expectancies of behaviour and not necessarily actual behaviour. In *Figure 13* below is a similar map, drawn to represent the actual social interaction of the members of the management group.

FIG. 13: *Interpersonal Interaction of Radbourne Mill Management*

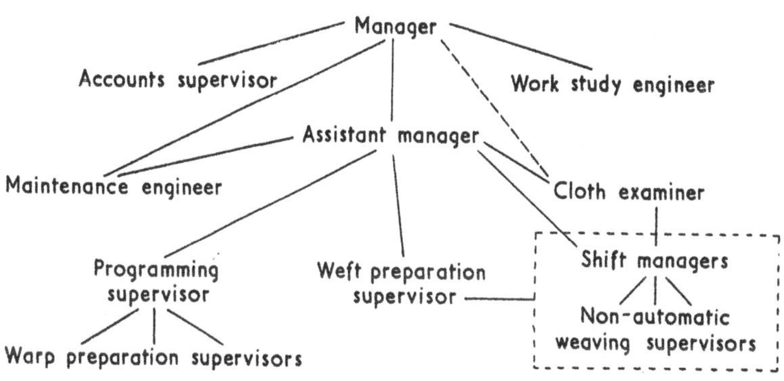

We are trying to indicate that, from the day-to-day working point of view, there were important sub-groups within the total management group itself. Primarily these were: (a) the *weaving* group centred on the shift managers and the assistant manager (himself a weaving technologist, former overlooker) and (b) the *warp preparation* group. We are not attempting to say that these groups were exclusive – which would be quite untrue – but that the groups were organized informally along these lines despite variation from individual to individual within each group. We substantiated this grouping both by observation of the supervisors' behaviour and also by using a technique developed by Burns (1954) which entailed the supervisors' making a record of their own interactions with each other. This study was carried out over one working week by all the supervisors and confirmed the groupings which we have indicated in *Figure 13*.

From this analysis, two things emerge quite clearly: the *relative* separation of warp preparation and weaving; and the mid-way position, closer to weaving than to warping, occupied by weft preparation. Historical as well as current functional factors contributed to this clustering: namely, the traditional separation between weaving and non-weaving departments. Weaving supervisors were always recruited from the ranks of the overlookers, who, it will be remembered, occupied a peculiar position of high status and power in the mill. Preparatory supervisors were recruited from outside the ranks of either weavers or overlookers. Hence, a sense of somewhat different identity was imported into the current situation. This is another historical factor which is difficult to reconcile with the production requirements of the automatic looms. The Debenham Mill organization miminizes the importance of this factor, but it is still present to some degree.

At this point it is pertinent to ask to what extent the formal social system overlapped with or matched the pattern of informal interaction that we discerned. In other words, were informal relationships supported by any formal means?

At the beginning of the innovation the only formal meeting was the foremen's monthly meeting, which was held in accordance with the personnel policy of the company. This meeting was presided over by the manager, and there was an elected secretary to record the minutes and deal with the business. Then, soon after the appointment of the shift managers, they and the manager and assistant manager began to meet together weekly to discuss the current work programme. Some time later, not long after Black had been given the joint appointment of warp-sizing supervisor, a daily meeting was started whose membership was the four managers, Black, and Lawson the spooling supervisor. It will be remembered that Black was appointed in January 1956, a year after the innovation had begun. This meeting and its membership – if we focus on the automatic section of the mill – is reminiscent of the daily management meeting at Debenham which was discussed in Chapter Three.

Since the foremen's meeting was held only monthly, we can hardly assume that it supported the informal relationship which we noted, and certainly not with respect to immediate production tasks. From the point of view which we are adopting at present, it served the important task of giving the 'management group' as a whole some actual support by enabling all the members of the group to meet together to discuss matters of mutual concern.

There were, however, two supervisors who did not attend this meeting because they were not considered full-status supervisors. These two were Jackson, the programming supervisor, and West, the yarn store supervisor. In West's case this is perhaps partially understandable since much of his information was held by Jackson, but Jackson's exclusion from this meeting is significant. We have already mentioned that the assistant manager had referred to Jackson as the 'third man' in the mill but, quite clearly, whatever informal rights Jackson may have had to this title, he was not formally recognized at all as being a member of the 'management group'.

The weekly meeting of the managers can best be understood in terms of the social structure as giving form to the senior management group within the mill. It will be noted that this meeting is based more on status than on function since no preparatory supervisor is present as such; and since the groups we have suggested on p. 83 are almost certainly functional rather than simply status groups, this meeting does not especially serve to support the informal group behaviour.

The last formal meeting was the one which started late on in our study, and included Black, Lawson, Laycock, Oakroyd, and either Gardner or Ashby depending upon which was on duty. This meeting was closer to the existing informal grouping in that one or more members of these informal groups were represented at this meeting. And in addition this was much more nearly a 'functional' or work group, and much less a 'structural' meeting than the other two. Of the three meetings this was one which could be regarded as being primarily determined by the new conditions under which the mill was now working, but even so there were difficulties. From the point of view of liaison between preparation and weaving, the meeting was desirable, but the considerable differences in status between the members of the group were unfortunate. This was especially so between the weaving and preparatory departments, and was perhaps heightened by the fact that the managers (all weaving specialists) met once a week on their own.

From comments made by the participants it was apparent that this meeting acted mostly as a communication link between the manager and selected supervisors. For instance, one member commented: 'I suppose really it's for the benefit of Mr Oakroyd. It enables him to get a good picture of what is going on and why looms are stopped and so on.' And another said: 'It's a help to me because you can get decisions on things that might otherwise take weeks to decide.' In this sense, the role of the supervisors in the meeting was not an executive one.

Thus, on examination, it seems that most of the scheduled meetings held in the mill reflected its formal structure and can thus be designated as 'structural' meetings. On the other hand, the informal interactions seemed to be more nearly grouped (inevitably) along 'functional' lines and in that sense cut across the formal social system within the management group. In particular, the anomalous position of Jackson, the programming supervisor, in relation to the functional and structural divisions within the management group emerges very clearly. This once again points up the dissonance between the older work culture and the newer evolving culture of automatic working.

SUMMING UP

In this chapter we have attempted to draw together the relevant changes in the management group, and the attitudes taken by some of the supervisors towards these changes. Presented formally, the organization at the end of the innovation is shown in *Figure 10*, p. 64, and the most striking thing about the structure is that, as a result of the technological changes, it became more complex in terms of the number of status levels – although we have also remarked that there is a considerable difference between a new status and role being created in an organization and the acceptance of both status and role by other members of the organization.[1] Difficulties were experienced both by the supervisors and by the workpeople because, in part, the holders of the newly created posts of shift manager and sizing-warping supervisor had conflicting roles to enact.

Taking the management group as a whole, we postulated that the impact of the technical innovation would be felt primarily in two fields – quantity of production and quality of production. Our evidence was that on the whole there was an adequate reservoir of skill to enable the supervisors to deal with the demand for increased quality, although there were differences between warp and weft preparation supervisors in awareness of quality demands. More important than the production of high-quality materials for automatic production, is probably the feedback of information about the quality as it is being woven into cloth. Here, too, we have noted organizational differences between warp and weft preparation which militate against effective communication.

Quantity of production was rather different in its effect on the

[1] Woodward (1957) has investigated the pattern of organization arising from different stages in industrial production in several industries, and Radbourne Mill could be cited as a special case in her general argument concerning the tendency to integration of function.

group, problems arising particularly from one aspect, that of coordination of the work involved. Here, the historical isolation of departments and differentiation of function have constituted a factor operating against the integration and cooperation which automatic techniques demand for efficient running. In some ways, the most basic differentiation is that between weaving and non-weaving, and we have been interested to see how the upper levels of the Radbourne hierarchy are composed solely of weaving specialists.

Our evidence concerning the informal relationships within the group pointed up this cleavage, with the assistant manager acting as a vital 'bridge person' between the two major divisions. However, no formal meeting, apart from the foremen's meeting for the whole group, either acknowledged the existing functional split or attempted to close it. The one meeting that might have been an exception to this (i.e. the daily meeting started in 1956) in fact seemed to serve a rather different purpose.

Fortunately, the supervisory group was a stable one and so the tensions arising out of the conflicting demands of the technology of the mill and the culture pattern of autonomy of action were tolerated by the group as a whole and by the individual personalities composing the group. The mill was still small enough for every supervisor to know every other supervisor in a face-to-face relationship, and this undoubtedly helped in the solution of some of the conflicts we have spoken about. The fact that only a part of the mill was re-equipped had its own specific disrupting effects, which affected both the group under consideration and the workers – a point to which we shall return in later chapters.

In conclusion, our evidence is that the whole management group was affected by the innovation, and at the end of our study the adjustment processes were still continuing. The close integration and simplicity of the Debenham system could not possibly be operated at Radbourne because of the increased size of unit, but other solutions to the same set of problems were possible and tended to follow the Debenham pattern – that is, integration of function rather than differentiation, and a move from competition to cooperation between the individual departments that formed the total production stream.

REFERENCES

BURNS, T. (1954). The directions of activity and communications in a departmental executive group. *Hum. Relat.*, vol. VII, pp. 73–97.

EMERY, F. E. & MAREK, J. (1962). Some socio-technical aspects of automation. *Hum. Relat.*, vol. XV, pp. 17–26.

WOODWARD, J. (1957). Control and communication – a management concept of cybernetics. *J. Inst. Prod. Engrs.*, vol. 36, pp. 539–48.

SIX

Changing Tasks for Operatives

Two main groups of operatives – the weavers and the overlookers – were directly affected by the introduction of the automatic looms. They, unlike many other operatives in the mill who were only partially affected, found that their whole work life was now different. In this chapter we attempt first to understand how the task and responsibility of these operatives changed with the innovation. However, the definition of such a change in role in a complex social system like that of Radbourne Mill does not mean that it was straightaway clearly perceived and adopted by the persons concerned. Many factors influence the attitude and behaviour of individuals and groups in such a situation and some of these will also be discussed.

THE MACHINES THEMSELVES

Automatic looms have been described as more intricate and precision-built than non-automatic looms. Their main technical features are a weft-replenishing mechanism, a warp-stop mechanism, and an automatic let-off.[1] The first of these was the dominant new feature in the present case. The second had been fitted for some time on the non-automatic looms in the mill and the third, though recognized and appreciated by both the weavers and the overlookers, was not experienced as a major or direct change. The new looms were two-box looms built to weave silk, rayon, and other man-made fibres in fabrics calling for a maximum of two shuttles.

[1] A mechanism for rolling up the woven cloth automatically.

These automatic looms incorporated a mechanism which automatically changed the spool in the shuttle to replenish the weft. Spools were stored in a magazine on the end of the loom, and the empty ones collected in a canvas bag below. In operation, the new looms made more noise than non-automatic looms and they ran at speeds 20 to 25 per cent higher. A non-automatic loom had to be stopped and started for every spool – at intervals of 1 to 20 minutes depending on the nature of the yarn – whereas, in theory at least, these looms once started need not stop at all. In practice, such continuous running was rare, owing to a variety of technical and human inadequacies, but some did often run for very long periods without a stoppage. After the nylon quality had been in the looms for some time, a number of them ran for the whole of an eight-hour shift. This was quite a common occurrence at Debenham Mill.

The looms certainly were much more intricate than the non-automatic looms, which appeared as bare frames beside them. Even the old shuttle-change automatic looms were much simpler and did not have suction holding of weft ends and centre weft forks.

WEAVERS UNDER TRANSITION

The introduction of the automatic looms meant changes in the role and responsibility of the weavers. Some of these changes were obvious; others were less so and were merely implied by the new character of the production. The work-task of an automatic weaver can be defined in terms of these changes. This definition can never be an absolute one, because much depends upon the perceptions of automatic weaving held by those responsible for organizing the new task. These human perceptions add to and subtract from the more immediate effects of changes in machines.

The first factor of importance for the new role is the sett of looms. The difference between weaving on one automatic loom and on one non-automatic loom indicates only some of the change in the weaver's role. A sett of looms is the weaver's domain of responsibility and sets the boundary of his task, so it is in terms of this limit that we must try to define the changes in his work.

Work-study methods had been used in the company to analyse the work-task of weavers into a number of components, which are given in *Table 7*.

TABLE 7 WORK-STUDY COMPONENTS OF WEAVING

Non-automatic Weaver (4–8 Looms)	*Automatic Weaver (16–24 Looms)*
Warp stoppages	Warp stoppages
Weft stoppages	Weft stoppages
Other stoppages	Other stoppages
Shuttle-changing	Battery-filling
Walking	Walking
Supervision	Supervision

This analysis attempted to describe quantitatively the work-task as it was broadly perceived by the management.

That the loom, either automatic or non-automatic, did stop because of weft, warp, or other failures was apparent from its operation. So also was the fact that shuttles had to be changed, in one case, and storage batteries of weft pirns filled, in the other. All these conditions involved manual operations on the part of the weaver which could be reduced to a fairly standard pattern and so observed and measured. It was also obvious that if there were more than two looms in a sett, the weaver would be involved in some walking, but the nature and amount of this component was not so clearly defined or measured. Finally, the component called supervision was used to cover, in a rather vague way, the fact that, because stoppages of looms could be forestalled and because bad quality cloth could be produced from looms that did not stop, these two aspects of the task could be assumed to take up some, though how much it was difficult to say, of a weaver's working-time.

The three stoppage components and shuttle-changing all brought the non-automatic weaver into direct contact with the process of weaving the cloth. They were components that involved manipulative dexterity and for which the weaver was trained for some time. These stoppage components were still present, though less frequently, for the automatic weaver. Now, however, his other manual component – battery-filling – was a simple repetitive task which was done independently of the weaving process[1].

Supervision was officially described as including inspection, patrolling, dealing with loose ends, etc. The time allowed for supervision was increased with the size of the sett. Moreover, the sensitivity of the automatic looms to deviations from perfect conditions of operation, and their proneness to certain sorts of damage while still running, made fulfilment of supervision by the weaver more essential. Supervision thus became an important part of the weaver's task.

The automatic weavers did not do any cleaning or oiling of their looms, but these components had also disappeared from the work-task of the non-automatic weavers at Radbourne Mill before the innovation. At Debenham Mill and in the non-automatic departments at Radbourne, ancillary operatives removed the packets of cloth from the looms. In the automatic section at Radbourne the management reversed their specialization policy for the weaver's task by including this packet-handling in his role.

Such a reversal in the general trend of the change of role was

[1] It is a common practice in the industry (though not in this company) to use ancillary operatives for magazine filling. Juveniles and other unskilled operatives are used in such a role at rates of pay considerably less than that of the weavers.

received, as might have been expected, with marked disapproval by the weavers.

> 'I can't understand why we have to do it. They don't on the non-automatics, yet many of them are men. We get no allowance for it, although it takes time and makes it much harder.'
>
> 'I don't know why it is. Perhaps it's a distinction between men and women, but I'm going to bring it up with someone.'
>
> 'I first met this in the autos. I waited for someone to come and remove it, but they didn't. I've spoken to the managers about it but they won't wear it. I took ten off yesterday and we don't get paid for it. On the other looms, they are carried off. I just can't understand them.'

The number of looms in a sett, and consequently the area of the weaving shed they covered, was higher for the automatic weaver, which increased the proportion of his work-task devoted to walking and further increased its supervision component.

To summarize the difference in the weaver's task, three main features stand out. First, there was a considerable reduction in the proportion of the work-task which was manual, required training, and was directly related to the production of cloth. Second, the simple physical components – walking and battery-filling – which did not bring direct contact with the cloth or the weaving process increased considerably. Finally, supervision – an ill-defined component – which involved observational and mental activity rather than manual, also increased considerably.

This difference can be more sharply seen in the theoretical case of loom and yarn perfection. With non-automatic looms the weaver would be largely engaged in carrying out the manual task of reshuttling the looms of his sett. This requires a dexterity which would be directly reflected in the quality and quantity of finished cloth. The automatic weaver would be manually involved with the weaving process only twice per working day – to start his looms and to stop them. His other manual operation would be battery-filling, which is simple and only indirectly affects the finished cloth. Even battery-filling, if in fact it is carried out by the weaver, involves him at each loom – the productive and traditional unit if not the operative unit – for only a fraction of the time needed for the non-automatic weaver, since a battery holds up to sixteen pirns, which will last from half an hour to several hours according to yarn thickness. Walking and supervision would thus occupy most of a weaver's working time in this hypothetical case of technological perfection.

At Radbourne Mill, such weaving was certainly only a theoretical case for most qualities, but, as mentioned already, some weavers did begin to experience something of this perfection with the nylon quality.

Imperfections in the yarn or in some aspect of the preparation processes, caused stoppages in both types of loom.

With the automatic looms, stoppage faults were easier to repair than with non-automatic ones, and the starting of the looms was also simpler. The innovation occasioned a general drive in the mill to improve the preparation of the yarn for the looms, and as this was gradually achieved, these stoppages occurred less frequently. The automatic weavers were also relieved of the task of coping with 'smashes' – major stoppages involving a number of warp ends. Helpers and spare weavers were placed in the section to cope with these time-consuming repairs which had originally been part of the non-automatic weaver's role.

The repair of these faults (and reshuttling for non-automatic weavers) was the part of the weaver's task for which training was provided and to which the term 'skill' was applied in the factory. The damage to the cloth which always ensued at such stoppages was minimized by the 'skill' of the weaver.

This use of the term 'skill' in the factory is of great importance if we are to understand the changing relationships, attitudes, and behaviour patterns which form our subject. This 'skill' is a complex concept. It includes psychophysical elements similar to those described by Conrad and Siddall (1953) in their study of other textile processes, e.g. pirn winding; but these are interrelated with many more dynamic factors arising from the whole social system. Hence while the difficulty of a particular work-task or component is important, its significance for the total process also contributes to the 'skill' assigned to it. This relates in turn to the responsibility attached to that particular role in the social structure. When applied to roles, 'skill' depends on the range of different operations involved and the totality of their effect on the product. Yet again, the training provided for the role and the corresponding patterns of reward and punishment all contribute to its 'skill'. Further, the word 'skill' is applied to individuals in particular roles, involving such elements as sex, personality, leadership qualities, and length of service. The 'skill' of a role contributes largely to its status and to the prestige of individuals who occupy that role.

Because the word 'skill' had these dynamic implications we have avoided its use so far in this discussion of the changing function of operatives. It is not the neutral word that it is in other psychological contexts. Groups and individuals in the mill used it in quite different senses and in ways which showed how charged it was with social meaning. Thus, while overlooking was agreed by the management to be the most 'skilled' operative role, there was great divergence as to the rating of other roles on this dimension. Individuals with experience of the industry in Northern England tended to rate weaving next highest, while local people often placed warping above weaving – a remnant of particular features of this mill's history reaching back more than fifty years.

OVERLOOKERS UNDER TRANSITION

The other main group of operatives affected by the innovation were the overlookers. When we attempt to define their work-task on the two types of loom, there are no standard components such as those listed in *Table 7* for the weavers.

Under the company-union agreement, no work study of overlooking was undertaken. However, the nature of the machines themselves clearly indicates some important elements of change in this role. An overlooker, both automatic and non-automatic, is essentially a maintenance mechanic or engineer. His task covers the preparation of the machines for production, and their adjustment and maintenance during production. Automatic looms, we have said, are much more complex machines and are built to more precise specifications. Thus, primarily, the overlooker had to master a more complex mechanism, for which, for the first time in the history of the industry, a printed manual was available. He had to develop a standardized approach to the looms in which all settings had to be made as prescribed by the manual, and to a new degree of accuracy. These settings, though more critical, were easier to make if the mechanism was understood.

These changes relate to the single loom; but, like the weaver, an overlooker was responsible for a sett of looms. Although there was little change in the size of their setts because of the innovation, some features of their changed role do relate to the sett of looms. These features are not clearly defined, but arise from the increased continuity of production with automatic looms, the more serious economic consequence of looms standing idle, and the increased size of the weavers' setts. The automatic overlooker had an increased supervisory function which covered both the stopped and the running looms of his sett. He could no longer rely on the weavers to notice and report all inadequacies of machine performance. His supervisory function was thus complementary to the new supervisory component of the weaver's task. The non-automatic overlooker had a role which was more discontinuous and which consisted of a series of tasks on stopped looms. These differences implied a distinct shift in the working relationship between weaver and overlooker.

To summarize these differences, the automatic role has acquired a more precisely defined manual component, and an extended supervisory component that carries implications for its relation to the weaver's role.

When we examine the perception of these changes by the persons connected with the section, we find that these two aspects are given varying emphasis. The shift managers and the overlookers themselves perceived initially only the manual components of the new role, which we would describe as a 'loom-centred' view.

> 'The overlookers are the key people, and to a man who has spent many years on the older types of looms, the new looms do present many new and different features. On the old looms there were settings and adjustments, but things were much more flexible and tolerant. The looms would still operate quite satisfactorily over quite a wide range of setting. Now, however, although an automatic loom could be run in that way, it is not the way to run it efficiently or satisfactorily.
>
> 'The old type of loom required a lot of initiative and individual playing about, to match loom and cloth for production, but for the new looms – although, of course, there is still a need to fit to particular yarn and cloth – the settings are known and mechanically these must be adhered to.
>
> 'For an old-type overlooker, the problem is largely one of forgetting what he already knows and being prepared to start afresh with the new material. The old loom overlooking was very much learnt and picked up on the loom. Now there is a manual, and it is a lovely job – all the information is available there.'
>
> 'There is an enormous difference between non-automatics and automatics. The automatics are more complicated and you have to follow the book. There's more to do, more parts to adjust and you are always on the go with these looms.'
>
> 'On the old looms you could do anything and get away with it, but there are fixed settings on automatics and they must be followed from the manual.'

In contrast to such comments are the following from Laycock, and from Knight, the overlooker from Debenham Mill, whose perceptions were in terms of the supervisory component – that is, a 'sett-centred' view.

'Automatic overlookers should be constantly on the alert about their looms. They should observe the running looms. Look at Young and Irwin – they are good. You never see them loafing round as you do the other overlookers in the non-automatics. When they finish a loom they watch the cloth, looking for work.'

'The thing you've got to reach is a state when you know just what's wrong and what needs doing as soon as the weaver comes to you. It's no good if you have to go to the loom and start looking for the trouble.'

These differing perceptions were consistent with the essentially 'non-automatic' and 'automatic' frames of reference held by these people respectively. Throughout the study, there was a gradual change in the local overlooker's view of their role. In particular, they were aware of increased manual continuity and of the need for some supervision of the sett on their part. In August and September of 1955 the following comments were typical ones made by the new overlookers:

'The automatic is very different from non-automatic looms. You have to be more precise and there are more complications which can require attention. You are more at it on autos.'

'There's more work on the automatic. You are really kept at it all the time. Before, I could say, "Well, I've an hour to spare". I never have it now, and I don't think even the experienced ones in the automatic looms can stop like we used to.'

'The thing is the weaver just hasn't time to see all the faults. He can go back and forth around 24 looms and not see anything wrong. Yet there may be a bad fault in the packet. The overlooker does have to keep the loom in the condition to weave good cloth. He, in turn, becomes more dependent on the weaver to see that the standard is up, or else they must report the loom to him.'

'The weaver hasn't time to see all the faults and the overlookers must keep an eye open to try and correct it. More is left to the overlooker with autos. You are more dependent on each other on the automatics. On non-automatics the weavers and overlookers were able to work more on their own.'

But, though, like the majority of the weavers, the overlookers changed their frames of reference as their experience of the new looms required, their attitude to their new role and to the looms remained more nearly a non-automatic one. One overlooker said, for instance: 'I don't mind the automatics. There isn't the same skill and craft required on them. Everything has to be done by the book. In a way, because they are more precision looms, they are easier to adjust. Before you had to know your yarn and adjust your loom to it. You couldn't become proficient in a couple of years before.'

THE OPERATIVES LOOK AT AND TRY TO ASSESS THE CHANGES

The new roles for the automatic weavers and the overlookers formally appeared in the social structure of the mill with the arrival of the new looms. However, the occupants of these new roles did not immediately take on the patterns of attitude and behaviour we have just defined. The gradual evolution of these new patterns depended upon an interaction of many factors. For some individuals the evolution was rapid and almost complete, whereas for others it was slow and incomplete. Virtually all were seen to experience it during the period of our study. An individual's perception of the changing roles depended on his attitude to the whole changing situation, and this in turn was affected by his experience of the changes as they occurred.

All the operatives when they first entered the section perceived the innovation within a frame of reference which was rooted in the non-automatic pattern of the mill to which they had been accustomed. In general, their attitude to the looms was unfavourable and found expression in contrasts of particular features of the two types, which usually stressed the disadvantages of the new.

'I liked the little non-automatics. You could handle them better. It helps to be tall on the automatics.'

'The extra braking is the thing I find most difficult. If you don't stop them just right, you have trouble starting up and there is time wasted. With the non-automatics, you just knew that they would run on for two picks. The tensioning is better in the automatics.'

'Yes, there is a big difference. One hundred and twenty-four picks per minute to what I've got now. Also, there was only a single change of shuttle. But they seem to be well built and designed.'

This 'loom-centred' perception was fostered by the fact that the first weeks or months an operative spent in the section were centred on the single loom and not on the sett of looms. The training programme had this emphasis, and furthermore the non-automatic pattern was in itself essentially 'loom-oriented'. The strength of this feature of the non-automatic pattern is borne out in the continuing difficulties that arose in the mill when the payment was changed from a 'per loom' basis to a 'per work-load' basis – work-load being the unit used by the work-study department.

The weavers who were recruited especially for the innovation also showed this loom-centred perception in their early days in the section, but their general attitude was freer of non-automatic traditions and preferences.

'There is a lot of difference between non-automatics and automatics. The automatic is faster, but for me, a new person, they are easier. The loom does everything for you. It stops on the spot. You don't have to push the loom about like the non-automatics. Filling the magazines couldn't be simpler. I learnt on non-automatics but found them hard to pick up.'

Only during this phase of the evolution of an operative's attitude in the innovation period is it at all possible to speak of his attitude towards a single automatic loom. Soon the operative found himself responsible for a sett of the automatic looms, and generally entered a new phase which can be described as 'work-life-centred'. The frame of reference was still essentially non-automatic for those who had experienced that form of production. However, now our questions about the nature of the differences between the looms no longer received the loom-centred answers of the first phase. Rather, features of the whole work-life of the individual, such as multiple-loom weaving, shift work, work study, the wage system, and relations with overlookers and ancillary workers, were mentioned.

> 'We still get a lot of trouble with the automatics. There's nothing like a non-automatic for good running. There's something about the automatics that keeps troubling you. I think I would prefer a full sett of non-automatics. You can get a system with them.'
>
> 'I didn't notice much difference between one loom and another, but I do now. I have to work much harder with a full sett. When helping there isn't much difference between one loom and another.'
>
> 'They are different to the others. The main thing for me at present is to learn to work opposite someone else on the other shift. That's the biggest difference from the non-automatics.'

Some features of the new experience were favourable and others were unfavourable, but the yardstick was still a non-automatic one. For some of the operatives perceptions and attitudes characteristic of this phase persisted throughout the entire study. For others, at varying points in their experience of the new section, a third phase appeared. In contrast to the other two, this phase was centred on an identification with the automatic section and meant that new automatic frames of reference had been developed. Most of the weavers can be said to have reached this third phase towards the end of the study. Their general satisfaction with their new role is illustrated by the following comments:

'I haven't thought about it. If we got the same money, I suppose I would prefer a smaller number of looms, but I wouldn't like variability. I quite like the auto loom. People are happier on autos.'

'No. I think I prefer the automatics. Automatic machinery is really better. There's nothing query, query about it. No string and wax holding it together and you can get on better with them.'

'I like the autos now – better than the non-autos. They seem to run better and give less trouble. There's less to do on the looms. No, I prefer 24 looms with less worry and trouble to 6 looms. I used to like the patterns but I don't miss them now.'

'No, I shouldn't like the non-automatics. The automatics are modern looms and it's better to be on new modern equipment.'

Among the new overlookers as a group, this automatic frame of reference was not as sharply developed.

'I think I like them now. It's been a good change for me. I'm a full man now.'

'In most ways, I prefer the non-automatic, but the change to autos is coming and it's better to change now than in six month's time. There's more work in the automatic. You are really kept at it all the time. On the non-automatic you could think, "Well, I've got an hour free now". I never have it now and I think even the experienced ones are kept at it.'

The four weavers and two overlookers who had been on the original 48 automatic looms clearly showed this identification with the automatic section from the time of its establishment in the main shed. Status and prestige of seniority was conferred on them by management and by the newcomers in the section. Furthermore:

'The automatics are the best looms we have in the factory.'

'There's a big difference – speed of running and the number of looms. I like automatics now. The number can put you off to start with but you get used to it.'

They were responsible for training many of the other operatives, and all these factors had an effect on their own attitudes and frames of reference.

> 'There are enormous differences between non-automatics and automatics. You have to forget all you know. It's not a help at all – in fact, it's a hindrance. I prefer the automatic to the non-automatics.'
>
> 'The automatic is very different from the old looms. You have to be more precise. I'm all in favour of standardization. There should be more of it in other sections like the non-automatics.'

ON THE JOB

We will now consider the extent to which the behavioural requirements of the new roles were assimilated by the operatives.

The Weavers in a New Role

A number of criteria were tried to assess the weavers' behaviour in their new role. Performance records existed for the weavers and a criterion such as they provided was desirable, since it could be used retrospectively and without our intrusion into the behaviour scene. Most of the weaving managers and the overlookers agreed that *some* combination of the quality and the quantity of his output would indicate the 'goodness' of a weaver's behaviour in the automatic role. However, no suggestion of what this combination should be was forthcoming. They commented:

> 'Well, quantity isn't the whole story. There is a tendency to be 'pick-conscious' and not to bother about cloth, but the best weavers do achieve both quantity and quality. Not many of our weavers supervise adequately . . . I can give you a particular weaver's ability, as I see them all the time.'
>
> 'There aren't really any figures you can take to give you that information, or at least they would have to be taken over such a long period and are so complicated that it would not be worth while. You can tell who are good weavers, but it is not

easy to get figures that would mean anything. . . . The weaving efficiency figure gives you about 85 per cent of the picture, but it is only a guide to the quantity of production. We have those figures going back over a long period, but they are not simply interpreted because of changes of quality in the middle of periods, etc. Efficiency could drop markedly but it would not necessarily mean bad weaving.'

'The grading of the packets, again, tells some of the story as far as quality goes, and quality was what I meant by the other 15 per cent, but sometimes we put good weavers on what we know will be difficult qualities and they may have a bad run In general the good quality on the board comes from certain weavers. . . . Most times, in good weaving, quality and quantity go hand-in-hand, but sometimes they oppose each other. If a loom keeps on running, which means high quantity, there is a chance of the cloth being good, i.e. the stoppage faults are being eliminated, and in this respect quality and quantity do go together. On the other hand, there are all the faults of the cloth which don't cause the loom to stop, and adequate supervision of these means interference by the weaver in the running of the loom, with a drop in quantity. In this case quantity and quality oppose.'

All attempts to use these output figures were thwarted by the variable nature of the weavers' conditions during the study – the presence of trainee weavers, experienced or inexperienced over-lookers, and different qualities in the setts.

The second criterion was the observed behaviour of the weavers in their automatic roles. We spent many hours observing the weavers in the automatic section, and particular attention was paid to the way the weaver moved around his sett of looms. Most weavers did develop a system for attending to the looms in their sett, although this was mainly related to the tasks of filling the batteries and attending to stopped looms. These two operations were pre-eminent for all the weavers. A few of the weavers regularly spent time observing the backs and fronts of running looms, but in general little time was spent on this type of pure supervision. Only in isolated cases were weavers observed to stop a running loom and report it to their overlooker. Nevertheless, it was only with the nylon quality in their looms that weavers were often observed to be idle and stationary for periods of longer than a minute.

The observations on the general lack of supervisory behaviour were repeated during the studies at Debenham Mill. In this case, however, several of the weavers behaved quite differently. These men and women – about 25 per cent of all the weavers – spent as much time per hour supervising and forestalling possible faults in the running looms as they did on battery-filling and attending to stoppages.

The third criterion was the assessment of the weavers by other people in the mill. The weaving managers, the cloth room foreman, and the overlookers gave their separate assessments. All these were unanimous in their high opinion of the four experienced weavers in the section, and this is consistent with the seniority and prestige that had been awarded to them from the beginning of the innovation, as described earlier. But consistent opinions about the other weavers were conspicuously absent. For example, two comments on weaver N.M. were:

> 'If you take N.M., for instance: she seems always to be by the loom that stops. Quite uncanny, but there it is. Week after week, she refutes the argument that you can't get both quantity and quality. Sometimes they do go against each other, but they can go together.'

Yet the cloth inspector said:

> 'Well, she is a good weaver, but I wouldn't call her one of the best. She is the sort who will let the cloth run on. Funny, we had her up last week. She's the sort, also, who can't be told anything. She does not believe that she's capable of producing bad cloth. There are those who say if you keep the looms

running the quality will be good, but it doesn't always follow, from my point of view. If you always have the picks in mind, you're not going to produce the best cloth. A good steady weaver is better.'

These diverging opinions are to be expected when the perception by these people of the role change itself is examined. The following comments reflect how these perceptions varied with the amount of understanding the person has of automatic weaving.

Oakroyd, the manager:

'It is a weaver's job to supervise, and he has to learn to remember and detect faults, as to whether they are purely random or due to some loom fault. This especially applies to non-stoppage faults. It is going to be a real problem with 20 looms, to remember where a fault was, its type, which loom it occurred on, and the regularity or otherwise of its occurrence, when a round of all the looms may take twenty minutes.... We are talking here about observing and supervising a circular sett of looms, i.e. when you can go steadily round from one to the next, etc. With different qualities, although I'm not saying that it isn't still better to treat it circularly, often the weavers don't work circularly and this makes remembering and checking even more difficult.'

Laycock, the assistant manager:

'There is a good automatic weaver at Debenham. He sticks to his work and spends his time at his looms. When not starting them up, he is anticipating faults and stoppages by supervision. He waits on his overlooker and when he is free, summons him to where he anticipates or has found a fault. He is shy of women and doesn't talk to the other men. The women don't supervise so well. They will stand and gossip.'

Gardner, a shift manager:

'First of all there is the general background of all the weaving: tying knots and keeping the warp straight. This is common to all weaving. Then there is organization – working out a system – which is much more important for automatic weaving than non-automatic. Starting-up is a great skill with older looms, but not so with the automatic. The automatics require less skill, and before, one used to speak of a good weaver being a 'good joiner-in'. The automatics need different requirements. As ideally run, with good warps and wefts, there would be little to do with an automatic, but at present a good weaver will supervise from the back and work systematically round the looms.'

Moore, the work-study engineer:

'I suppose it is one who produces good cloth, and one who keeps his looms running. Good weavers don't fly round their looms – they work out some system of working. A good weaver will call the overlooker early and won't run the loom on.'

Boyd, the cloth room foreman:

'It's a weaver we don't see here very much. It's a funny thing, but heredity seems to play a part. If a weaver's mother was a good weaver we often find she will be good – or again, bad weavers follow bad weavers. I don't know what it is. . . . Most of the automatic weavers are men and they don't make such good weavers as women. . . . Yes, it's possible that some of the men, who weren't good on non-automatics, may be better on automatics.

'These people aren't really weavers in the same sense. . . . There are women, who are real good weavers in the other departments.

'Auto-weavers are more overloaded. They haven't time to do the same fine job. They aren't as skilful, but they have to keep going at all costs. It's a dead-end job for men.'

Examination of these comments reveals the following points. The overlookers and the cloth room foreman all appeared to use non-automatic frames of reference in their perceptions and evaluations of the new role. Manual 'skill' on the cloth was a dominant concept and the increased non-manual components received little attention. Only Oakroyd and Laycock suggested that supervision is 'skilful' and hence status-conferring. So the status of the weaver's role, being defined by other operatives and by the rest of management in terms of traditional 'skill', was definitely diminished by the innovation.

All the comments (except those of the cloth room foreman, which were wholly quality-oriented – the terms of his own role) indicate an awareness of the need for a systematic method of covering the sett of looms. Nevertheless, except by the two top managers, there was no formal differentiation of the automatic role from the non-automatic one. Rather, the new role was seen as a degradation of the non-automatic weaver's role. 'These people aren't weavers any more – they are just handle-pullers and machine-minders.'

There was thus a failure by most of the personnel in contact with the weavers to recognize the totality of the change in the weaver's function. The effects of this failure on the way in which the weavers themselves reacted to their new role will be discussed later in this chapter.

Laycock and Moore refer to the relationship between the weaver and the overlooker, and this social aspect of the new role will be taken up again in Chapter Eight.

Finally, we obtained the weavers' own perceptions of their new role. These comments from the weavers reflect quite accurately their behaviour as we observed it. The figure given after the comment is the weaver's average efficiency when only the nylon quality was in the looms. It is really no measure of his 'goodness' as an automatic weaver except in conjunction with other factors, unfortunately indeterminate.

The original experienced automatic weavers said:

'You must go to the back of the looms as well as the front.

At the back, you can save a lot of time by correcting the loom before it stops. Also, if the selvedge is wrong you must find out why it is breaking. It's no good just to keep on putting ends in. A.B. fails to get good efficiencies for these reasons. We get 100–102, but he only gets 96–98. . . . A system is most important. I tell trainees to watch me working. I check the looms and fill the magazines in order.' (93·7)

'You must finish one job before you start another. The trouble is people won't do it, and they panic when they just leave a loom and find it's stopped again. Some would be good weavers if they kept calm. . . . You can't get on without a system. You have to check the looms as they are running. Above all you mustn't worry or panic.' (94·0)

These weavers did perceive and fulfil the automatic role more completely than the rest of the group.

The weavers from the specialist non-automatic department, which was associated with the highest weaving 'skill' in the view of most people in the mill, did not perceive supervision or systematic working as distinctive components of their role. Nevertheless, their actual behaviour led to tidier warps and fewer stoppages than many of the other weavers achieved.

'You have to keep going all the time. Much more moving about to do. . . . I don't think I have a system. If a magazine was empty near me, I'd fill it; but otherwise I just keep going to stopped looms.' (94·7)

'I don't have any system. I try and go round the batteries at the same time each day. If a loom stops, I go to it and then back to where I had got to with the batteries.' (91·7)

The weavers from the non-specialist non-automatic weaving department did in general perceive clearly the multi-loom feature of their new role, but only in terms of the manual components. They had had experience of setts of up to 12 looms in their former departments. They commented:

'Keep at it. Even when they are running well, you must keep on the move around them. I don't think a system is possible.' (94·4)

'You just keep going round them.... Yes, you soon work out a method of filling batteries, taking packets off, and going round them.' (92·1)

'I just keep going to each loom that is stopped. If they are all going, I fill up the batteries and attend to anything else that is to hand.... I don't see how you can really have a system.' (95·7)

'You must have your looms in good order, and that means a good overlooker. You have to work with him.... Yes, I do have a system; I stand in the middle and go to the loom that stops, fill the batteries near there, and then return, checking the other looms on the way. But you have to alter your system to fit the qualities. I may sometimes stand in the corner of the sett instead of centre, if I have certain qualities there.' (89·9)

'Keep calm and cool. Once you are on one loom, forget the others, even if eight are stopped.... I don't think I do have a system. I just get to the next stopped loom by the shortest route.' (93·7)

The weavers from the weaving school said:

'You have to have a routine. The shoot [weft] I try and do last, and never when a loom is stopped. If two are stopped, I do the shortest job first.' (95·6)

'Yes, my system is to try and fill all the magazines in order, and work round the back of the looms when they are running. That's something no one ever told me – to look at the back of the looms. I learnt that by experience.' (89·4)

The development of some form of systematic working was achieved by all the weavers, even though it was not always perceived as such. However, the system was almost invariably associated with the manual components of the task, which dominated the perception and behaviour of the weavers as a group. The non-manual components were perceived by some of the weavers, but in all cases there was considerable discrepancy between the time they devoted to them and that which would lead to optimum production, especially in terms of quality.

Several features of the factory as a social system were also related to these changing attitudes and behaviour patterns, and particularly to the discrepancy between favourable attitudes and good performance.

One has already been mentioned. A number of people in direct contact with the automatic weavers failed to appreciate their changed roles and to develop new cultural modes of behaviour appropriate to the new relationship between them. This was directly manifest to the the weavers in a number of ways. Punishment or sanction for their work on the looms was a striking example.

The behaviour pattern for such sanctions which existed before the innovation, was essentially as follows. Although, as at Debenham Mill, no financial sanction was applied, the weaver of a damaged packet was always called to the cloth room, where he was reprimanded by one of the managers, his own foreman, or the cloth room foreman. Sometimes the overlooker would also be reprimanded in this way. This traditional pattern for applying sanctions was unaltered for the weavers in the new section.

This occurred despite the emphasis laid by Laycock and Moore on the overlooker–weaver relationship, and despite Laycock's point-of-view expressed as follows: 'What can an auto weaver do about quality? The loom will start in any position and if it keeps running it produces better cloth than non-automatics, and keeping the looms running puts the picks on.'

In contrast to this view, Oakroyd assessed where responsibility lay in a way which he believed could be applied to both types of weaving, while recognizing that there were basic differences in the roles: 'I always hold and try to explain that the weaver is responsible for the cloth. The weaver is not responsible for the loom going wrong, but he is responsible for carrying on when it does go wrong. It is then the overlooker's responsibility to see that the loom works properly and doesn't stop.'

There was little awareness among the management, apart from the top management, that the quality of the cloth in the automatic section depended on a complex network of factory personnel.

This tradition of holding the weaver responsible for all damaged cloth from his looms was criticized in the Works Council meetings. Although the topic of damage was discussed at most of the monthly meetings during the period of the study, never was it suggested that the overlooker had responsibility for the damage as well. This is even more striking when it is remembered that the overlookers were not represented at these meetings. The only displacement of attention was that shown by the weavers on the Council in the scapegoating of ancillary workers of low status such as the oilers and cleaners.

There was no direct reward for good quality, though the payment system is expressed in such a way that, in theory, good quality cloth is all that is paid for. For example, a weaver said: 'We only hear about the bad packets. What happens to all the rest?'

This 'non-automatic' view of responsibility was experienced by the automatic weavers when they were repeatedly called from their looms to the cloth room to see damaged cloth. One of the best and most experienced of the weavers commented: 'Some of the foremen and managers here are gentlemen. Others, it's no good talking to them. I've been called up to the gauze room to see Mr Laycock – I'm no better than the rest of them. He runs through my packet and says "You've made a bloody mess of that, haven't you – see you do better with the next one". I've had a foreman, though, who comes to you and says, "Didn't you see that [a particular fault] there?" I ask you, with 24 looms, how can you see everything? Mr Laycock's O.K. You can understand him and you don't mind what he says. The other one and others like him are fifty years out of date. They still think in the way you could with 2 looms per weaver. They haven't caught up with mass production or automatic machinery.

'With 2 looms per weaver you could see every inch of the cloth. If they expect that with autos, you'll get an efficiency of 30 per cent, which defeats the whole idea of autos. What you should hope for is a high efficiency of reasonable quality cloth.'

It is interesting to note that, technically, the particular foreman referred to here was one of the most experienced on automatic looms, and was referred to by one of the managers as, 'the only automatically-minded one we have here'. His high technical ability, which was the criterion for the management assessment, was not perceived by the operatives in his behaviour towards them. In fact, from his behaviour, they inferred that he had a 'non-automatic', or inappropriate, attitude towards them in the automatic situation.

Others saw the contradiction between this mode of sanction and the exhortation from the management that looms should be constantly supervised. As one said: 'What's the good of saying the looms should be kept going all the time and constantly supervised, if we are called away from them for minutes at a time every couple of days.'

It seems that the maintenance of a mode of punishment from the non-automatic pattern retarded the assimilation of the new role and its associated attitudes. This hypothesis about the influence of inappropriate punishment could well be further investigated in other field studies or in laboratory experiments on groups.

The Overlookers on the Job

We also assessed in various ways the actual behaviour of the overlookers in their new roles.

Again, there were no simple quantitative records that could be used for this purpose. The efficiency of his looms and the quality of the cloth do reflect an overlooker's efficiency in his role; but in practice many other factors prevent such simple comparisons.

Observations of the group confirmed that only the three experienced overlookers spent any appreciable time supervising running looms. Such supervision was observed to be a considerable part of the total behaviour of a number of the overlookers at Debenham Mill. The new overlookers in the section at Radbourne were occupied with manual tasks on their looms almost continuously. In contrast to these two modes of behaviour in the automatic section, the overlookers in the adjacent non-automatic departments interspersed their manual activities on a loom with unoccupied periods when they would sit on a stool or in some other manner be effectively out of touch with their sett of looms.

The assessment of the overlookers by other personnel supported the observations. The management throughout the study expressed the opinion that the original overlookers were the only ones who worked in the appropriate way for automatic production. The slowness with which the others adopted the role led to the special training schemes and assistance which was provided for the overlookers.

A number of the weavers also commented on the failure of the new overlookers to be 'sett-centred': 'The trouble is not what they do. That is all right, but they don't work fast enough. With automatic looms you need to speed up.'

NEW GROUPS OR OLD

One factor which may have considerable bearing on the situations that have been described so far in this chapter is the nature of the groups of operatives concerned. In this study we are particularly interested in what we have called in Chapter One the 'cohesiveness of the group'. The individuals who made up the groups of weavers and overlookers, did so by virtue of the fact that in their working life they occupied these particular roles in the social structure of this factory. When they entered the employment of the company they accepted (during their hours in the factory) imposed conditions. They worked on particular machines in particular areas of the factory, which were prescribed for them by the management of the mill. The individuals – in the same role on the adjacent sett and in other roles – with whom they had contact were also rarely of their own choosing, but were chosen by the management. The particular managers who shaped their working destiny in these ways were also not chosen by the operatives themselves. Nevertheless, provided these foremen and managers did occupy the appropriate authority positions in the formal structure of the mill, this patterning of their work-life was accepted by the operatives.

In this way, we see that such groups as the overlookers, the weavers, the foremen, or any other role group in the factory, have essential differences from groups like a club, a neighbourhood, or an interest group. In these the members come together voluntarily, throw up their own leaders and together seek aims which reflect the interests of the individual members. It is to these 'free' or 'peer' groups, or to the similar artificial groups created by psychological and sociological investigators, that the term 'group cohesiveness' has been applied. Nevertheless, the concept is capable of extension to the 'role' groups of the type that exist in Radbourne Mill. Such groups, although formally determined, have nevertheless some important elements of freedom of behaviour which help to define the culture or way of life of the factory. For example, at Radbourne the overlookers all belonged to the Overlookers' Union – a group-determined aspect of the culture. Again, the weavers were free to join the T. & G.W.U. if they so desired. The individuals could choose whom they sat with for meals in the factory canteen, and whom they talked with most among the other operatives.

Many such optional behaviour patterns were open to operatives in all roles. The greater the uniformity that members exhibited in such patterns of behaviour, the more sharply defined was the group. In turn, the more definite was the group the greater were the pressures on individuals against deviation from the established patterns. In other words, group cohesiveness was greater. Not only can the members by their behaviour and attitudes increase the distinctiveness of the group, but the management and other groups in the social structure can do this also. By special conditions of pay, special patterns of punishment, or special patterns of inter-group behaviour, the management could help the group to become more definite and more cohesive.

After these general remarks on the cohesiveness of role groups in social systems such as factories, we will now consider more specifically the weavers and overlookers of the automatic section at Radbourne.

The Weavers as a Group

The weavers as a whole in the factory were an ill-defined group without any special union of their own. Some of them belonged to the T. & G.W.U., of which the official representative was a non-weaver. Conditions varied considerably in the different weaving departments. There was no standard form of training or apprenticeship for weavers. For several years both men and women had occupied the role, which previously had been for women only. Weavers were not infrequently moved from one weaving department to another or even transferred to other jobs altogether.

With the establishment of the new automatic section the weavers in it began to appear as a definite and more cohesive sub-group within the total group. First, weaving on automatic looms was sufficiently different from weaving on non-automatics for the weavers in this section to be perceived by many others in the factory as now being in a different position from the other weavers.

The managers in their turn helped to establish a new character for this group of weavers. Their policy of choosing only men for this type of weaving – even if relaxed later by necessity – gave the group a distinctive uniformity. Their special methods and, in particular, the fact that this group of weavers had higher potential wages than any of the non-automatic weavers both defined the group and afforded it important elements for developing a high status for its role. Even the management decision that these weavers should carry their own packets of cloth was definitive, and the weavers' comments suggest that they accepted it with only verbal grumblings because it was consistent with the 'maleness' of the newly emerging group.

All the weavers in the new section were union members, and as a group they did actively use this official channel for negotiating with the management. In the other departments these negotiations were more often between managers and individuals; or the union would take up an individual's case rather than that of the group of weavers.

On other occasions, described in Chapter Four, the automatic weavers insisted on uniform wage conditions throughout the group. In these negotiations the management on several occasions dealt directly with the weavers themselves, virtually by-passing the union representative. This recognition of the automatic weavers as a group distinct from the general group of weavers represented in the T. & G.W.U. was parallelled by a reaction from the group itself.

The suggestion was made in the section that the T. & G.W.U. representative, being a non-weaver, was unable to appreciate or adequately represent the group. Such ambivalence towards the union is to be expected in the dynamic situation of establishing the group.

In the patterns of work behaviour of the automatic weavers there was also evidence of greater uniformity and of the exercise of restraint, which supports the claim above that the cohesiveness in the group increased because of the innovation. One example was given in Chapter Four where the 'hooter' campaign has been described, in which the management again singled out the automatic weavers from the other weavers. Another example was the Works Council where, prior to the innovation, the few automatic weavers were officially represented by one of the non-automatic weavers. At the first election after the arrival of the new looms, one of the automatic weavers succeeded to this representative position, although the automatic weavers were still a minority. Thereafter, he vigorously represented the interests of the automatic section, further emphasizing to the representatives of the rest of the mill the special character of this new group and its role. The members of the automatic section, both weavers and overlookers, also began to confer as a group with the management over the operating conditions in the section.

A new note of urgency arose from their appreciation of the increased continuity of production and of its implications for their role. This led one of the shift managers to make the following comment:

> 'There is a real difference between the need for urgency in keeping the looms full. Automatic looms must be – others need not be so. We have more pressure on us from above to see that it happens, but also from below. The weavers and the overlookers on the autos have a much greater sense of urgency and they come to us if their looms are idle too long. On the non-automatics you can have a loom empty for a week and no one complains.'

Despite some initial resistance, the automatic weavers did develop a way of responding to this increased pressure which was uniformly adopted by them as a group.

In the non-automatic departments, some individual weavers came in early before the official starting times and, in this way, they obtained additional picks from their looms. The high efficiencies of these weavers were held up as examples to the automatic weavers by some of the foremen, in their endeavour to achieve more continuous production and higher efficiencies in the section.

Initially, some of the automatic weavers continued this practice after the section was established in the main shed; but soon after the 'hooter' incident, the practice was discontinued by all the weavers in the section. In this decision the weavers were supported by the overlookers.

> 'They quote these other people on non-automatics, who get efficiencies the looms can't give. They get them only by coming

in at ten to six or a quarter to six and starting up. Then the price goes down on our bonus. It's happened three times, but now we've rightly refused to start before six.'

'They've been on to us, saying it's the overlookers' job to see they [the weavers] keep running, because our wages will be higher. But if they think that I'm going to earn more out of other people working overtime, they'll soon find me in another job.'

This new norm of behaviour was impressed on each new weaver entering the section and was maintained quite uniformly throughout the study.

'I don't know whether I will be better off in the autos. We have been doing pretty well on the morning shift in the non-auto's, but you'll know about that. We get in ten minutes early, but on the autos they don't start till six.'

'We never start before six. On the non-automatics they do. We used to do so, but they [the group] decided not to, and we just never do now. On the other shift I think they start the motors at one minute to six, so that all of them are running at six.'

We suggest that this increase in the cohesiveness of the automatic weavers as a group was a dominant factor in the impact of the innovation upon them, in that their increased sense of belonging together as a distinct group facilitated the development of 'automatic' frames of reference and favourable attitudes to their new role. Support for this hypothesis is available from other types of group, such as those described in the attitude studies of Lewin (1952).[1]

[1] Another study in which the groups are formally constituted as in the present case, is that cited by Stouffer (1949) when it was found that racial prejudice among mixed army groups during the war subsided when a sense of belonging together in a common task – greater cohesiveness, in our terms – was achieved. In this study the reversibility of this relation was also demonstrated, namely loss of cohesiveness leading to renewed prejudice.

If we now consider the weavers who were the slowest to achieve automatic frames of reference, we find, as would be predicted by the hypothesis, that they were isolates (in various senses) in the group. One had a background of cultural and intellectual pursuits which gave him little common ground with the others, another was generally suspicious of any close relationships. He, in particular, had been a very good weaver in the non-automatic department, but in the new role an individual's superiority was not so apparent. This loss of personal prestige increased his antagonism towards the innovation and did not help him to overcome his natural isolation in the group.

A feature of this analysis of the weavers is the absence of any relation between satisfaction of individual weavers with the innovation and their output. This can be explained in terms of the development of the cohesiveness of the group and the changes in the weaving role. To achieve high performance in the role, we have seen that the weaver must spend considerable periods supervising his sett when all the looms are running. This means that the weaver is anchored within the physical bounds of his sett. Such behaviour is not conducive to the strengthening of the inter-member relationships in the group that would form part of the process of increasing its cohesion.

So we have the behaviour which helped to develop favourable perceptions and attitudes to the new role being at the same time behaviour which detracts from performance.

On the other hand, the comparative isolates were attendant on the looms of their sett more continuously and this gave them good performance, despite their generally unfavourable attitude and incomplete perception. These somewhat paradoxical results may be of considerable importance in selecting operatives for similar roles in which continuity and isolation are increased.

The Overlookers as a Group

Turning now to a consideration of all the overlookers in the mill prior to the innovation, we find, as compared with the weavers, a very cohesive group with norms, attitudes, and conditions which applied to all its members in the various weaving departments. All the occupants of this role were men and all were members of their own specific overlookers' union. The union had agreements with the company about the selection and training of the overlookers in the mill. The overlookers were uniformly paid at a considerably higher rate than any other operatives in the mill and their role had the highest status of all operative roles. This was recognized both by the management and the operatives, and was supported by the fact that all promotions to the weaving management were from the ranks of the overlookers.

The overlookers as a group had stood out against work-study and they also did not fill the seat allotted to them on the Works Council. Even the allotment of this seat to the group of overlookers indicates its strength, because the other seats on the Works Council were allotted on a departmental rather than a role basis.

The group was internally led by their elected union representatives, who during the period of the study were overlookers in the non-automatic weaving departments.

The innovation meant that eight of these overlookers were brought together in the new section of the mill. However, unlike the weavers, the overlookers did not experience this as an intensification of group sense. The new sub-group was not clearly more cohesive than the whole. On the hypothesis we have developed above, the pressures towards the acquisition of automatic frames of reference would be less intense for the new overlookers in the section. The relatively slow development of appropriate attitudes and behaviour on the part of the overlookers as described earlier in this chapter bears this out.

Nevertheless, there were practices among these automatic overlookers which indicate that they were beginning to develop cultural norms specific to their own situation, which were not shared by the larger group. For example, the two original overlookers in the section had, from its beginning, pooled the efficiencies from their respective shifts to give them equal wages. All the other pairs of overlookers in the section have followed this practice, although it was not general in the non-automatic departments on shifts.

The management did treat the automatic overlookers in some ways as a distinct group, although to a lesser extent than they did the weavers in the section. Thus, while there was not the same scope for direct and special negotiations over wage adjustments, they were given special training for their new job. Balancing this factor towards the formation of a new group was the attitude of the non-automatic overlookers, namely that there were not the same 'skill' and craftsmanship in the new role.

OTHER FACTORS IN THE INNOVATION

Some members of the management group expected factors rather different from those discussed above to be important during the innovation. We shall now consider some of these, and also an early attempt by the management to induce favourable attitudes to the changes in the operatives.

The automatic looms are more noisy than the non-automatics, and this feature was expected by some of the management to be a focus for an unfavourable reaction by the operatives. Except as a hindrance to instruction during training, this feature was never mentioned by the operatives. Such a contrary assessment illustrates the very different perceptions of the same situation that can arise even in so compact a social system as our factory.

Such differences in perception were again evident in the attempts made by the management to explain the change to the operatives before it occurred. The management, in line with the company's belief that the operatives should be kept informed about progress and development, announced to all the operatives the broad details and reasons for the change, as has been described in Chapter Four. Apart from the general good relations between operatives and management which were fostered by this policy in this case, there was little effect on the specific attitudes of individuals. The atmosphere of the mill was certainly one of confidence in the management fostered by the long tradition of good relations; but individuals still had to achieve new frames of reference with which to view and cope with the new experiences that would accompany the innovation.

When we consider these two aspects of the operatives' attitude to company policy – the one relating to general good relations, the other to the specific question of the innovation – we find that the management's efforts at persuasion were only effective for the first. The reasons given for the change were company-oriented – economic survival and the possibility of the decline of the town, by analogy with Lavenham. But the limited effectiveness of imposed verbal arguments in changing attitudes has often been demonstrated. Nevertheless, the publicity might have been influential in fostering a more favourable specific attitude had the case for innovation been presented, at least in part, in terms which were immediate and real for the operatives. Further, information about likely changes in the nature of the work itself might have supported and eased the operative's own learning by experience when the change began, and have hastened the evolution of new and appropriate frames of reference.

Finally, the multi-loom setts that would arise with the innovation were viewed by some of the management as another possible focus of disfavour, because of what they described as 'the weaver's attachment to his particular looms as unique machines'. In these terms, this concept may have had reality for some of the older weavers in the non-automatic departments at Radbourne. However, no evidence appeared for it among the operatives in the automatic section. Multi-loom setts *were* a focus of disfavour, but for other reasons, and in Chapter Eight the case of weavers sharing a sett of looms will be taken up.

The gradually increasing sett size and the introduction of shifts, together with the differentiation of the weavers' role, would appear to have done much to eliminate such emotional bonds between a weaver and his loom.

RESISTANCE OR SATISFACTION

As we have described it in this chapter, the technological change has meant for the operatives the evolution of new frames of reference. This was essentially the process so often referred to by the management as the 'overcoming of the operative resistance to change'. 'The management are always ahead of the workers in this sort of thing. There is a natural lag of the workers behind the management, and so such resistance is inevitable.'

This expression 'resistance to change' is commonly used in industrial situations at the present time, and it is one which is not helpful to relations between management and labour. It implies value judgements which are strongly resented in labour circles.

Technological change is almost invariably decided upon and initiated by representatives of management. In the process of decision, management develop frames of reference for the change. At the point of innovation these are already fairly developed and are of necessity rather different from those of the operatives, who at that time have no experience of the new technology.

What is experienced by management as resistance from the operatives is often merely the only response the operatives can make with their now inappropriate frames of reference. Further, what then appears to management as an overcoming of this resistance is the appearance of more appropriate responses on the part of operatives as new frames of reference evolve.

For the operatives, these responses are throughout consistently those best suited to maintain or improve their own position and satisfaction. This does not mean that real resistance to change never occurs, for it may not be possible for any new responses to appear that will maintain the satisfaction of the individuals or groups. *However, in a number of cases, including the present one, it seems that an inability to make appropriate responses immediately is interpreted as an unwillingness to respond.*

REFERENCES

CONRAD, R. & SIDDALL, J. G. (1953). An experimental study of pirn-winding. The effect of mixed setts on operative efficiency. *J. Textile Inst.*, vol. 44, pp. 215–22.

LEWIN, K. (1952). *Field theory in social science.* London: Tavistock Publications.

STOUFFER, S. (1949). *Studies in social psychology in World War II.* Princeton, N.J.: Princeton University Press.

Works not specifically referred to in the text

CROSSMAN, E. R. F. W. (1960). *Automation and skill.* London: H.M.S.O.

MAREK, J. (1962). Effects of automation in an actual-control work situation. London: Tavistock Institute of Human Relations. Doc. 669.

SEVEN

Selection and Training for Innovation

The introduction of new machines or any other technological change in a factory always leads to new roles with new functions. In filling these new roles, a company is faced with two alternatives. It can introduce new personnel, already experienced in the new functions, or it can select some of its existing employees to change their roles to the new ones that are required. If the new technology replaces an existing one, the prospect of redundant workers is an incentive for the company to choose the latter policy. In the first case, the company has no training problems, but there are the relational ones of settling the new people into the social system – particularly because these people come in to occupy roles that are also new. In the second case, there are two decisions confronting management. First, they have to select certain people from the existing social structure to fill the new roles with their new functions. Second, they have to decide on what form of training to provide for those they select.

In the present case, the company almost exclusively chose the latter course. That is, they selected certain of their existing personnel and set about training them for the functions of the new role.

In Chapter Six we have seen how the operative's function changed because of the innovation. We also examined how the perception and assimilation of these new functions was affected by the interplay of factors arising from the pattern of the mill.

In this chapter we shall describe how the company, in the midst of this dynamic situation, dealt with the selection and training of operatives for the automatic section. Once more, our interest lies in the interplay of underlying factors as they appear in these two aspects of the mill's life during the innovation period. Neither the training programme nor the selection criteria can be considered by themselves. They are part of the social system, and like all its parts, they are interdependent with the other areas. Our analysis of training and selection is thus related to the social structure of the mill, its culture, and the personalities of the individuals comprised in it.

Only through an understanding of how these factors interact can any predictions be made about the most effective selection criteria and training methods. The study of the present company provides material for this understanding, and an extension of such studies in other similar innovations could be of great practical and theoretical importance.

THE PLACE OF TRAINING IN THE COMPANY

There had been a tradition of a training school for weavers at Radbourne for many years. Watson, the sizing foreman in July 1954, had been in charge of it intermittently from 1924 to 1952. It had been leisurely in nature; and a new weaver would often spend a year or more in the school before beginning production work.

At Debenham Mill in 1949, the company had begun a more ambitious training scheme to supply the personnel for the post-war developments of shift work and automatic innovation. Until its closure in 1954, this school was in the charge of Bowyer, who was one of the foremen in the specialized non-automatic weaving departments at Radbourne during the present study. The Debenham programme was planned to prepare weavers for production in four months, and it also had some facilities for the initial training of overlookers. From 1952 to 1954 operatives from Radbourne were trained in this school, but Oakroyd was keen to have training facilities at Radbourne again, once the innovation began there.

Accordingly, at the beginning of 1954, a group of non-automatic looms in the N.E. corner of the main shed were constituted as a new weaving school, in charge of which was an experienced woman weaver who had helped Bowyer at Debenham. As mentioned in Chapter Five, this arrangement was used for training weavers throughout the innovation period until August 1955. Training was then virtually suspended until the beginning of 1956, when Watson left the sizing department and began another training school in a separate part of the mill, shown in *Figure 8*. Watson and Bowyer – both experienced in training weavers – took no part in the training programmes that are the subject of this chapter.

SELECTION OF OPERATIVES FOR WEAVING TRAINING

Until quite recently, all the weavers in the company were women and their entry to this role took the following pattern. Girls who joined the company began in unskilled jobs in the various departments of the mill. They became messenger girls and helpers in the spooling, sizing, warping, and examining departments. As the need for weavers arose periodically, some of the girls were drafted into the weaving school by arrangements between the weaving foreman and the departmental foreman under whom they were working. There was some competition between these foremen over which girls should enter the weaving school, since both naturally sought the skills of the most promising girls. This led to quite a traffic of 'misfits'. On the whole, after those entering the warping department, with its traditionally high standard of operative, those entering the weaving school had the higher intelligence and the greater manual ability. No formal testing of these abilities was carried out, and the selection was made on the basis of such general impressions as the foremen were able to gain. Some of those selected failed to make the grade in the training school and they would be returned to other jobs in the mill. No critical assessment of this pattern seems to have been undertaken at any time, although both Bowyer and Watson claimed that they could assess the ultimate outcome of the training in the first days or weeks after an operative entered the school.

With the advent of men as weavers in the post-war years, this pattern changed and the selection of prospective weavers was more often made by the top management of the mill. Specific recruitment of men for weaving replaced general recruitment followed by a later selection. There were not the 'unskilled' jobs for men and there was a new note of urgency in the recruiting. Finally, many of the men were adult when they were recruited, whereas in the old pattern recruitment was largely of juveniles. The managers interviewed the men who applied for jobs in a general way and again did not attempt any form of aptitude selection. Although there have been a number of aptitude tests suggested for weavers within the industry, both the managers and the experienced trainers believed that such tests were impracticable. They based this opinion on the grounds that the tests were too selective and that many people, who in time made quite adequate weavers, would not be selected. They also criticized the absence of any motivational elements in the test. 'With an ordinary person who is keen to learn, you can train them.'

Most of the men who applied to Radbourne Mill during the study were accepted and went straight into the weaving school. As at the Debenham school, the trainers assessed their progress after a month in the school, and some individuals were then taken out of the school by the management.

SELECTION FOR THE AUTOMATIC SECTION

If we now consider the particular situation that arose with the innovation of the automatic looms, two general criteria of selection for automatic weaving were used by the management. First, there was a definite policy of seeking *men* as weavers for the new machines. Second, juveniles and young operatives (under 40) were chosen to be trained for the new roles.

We will consider the consequences of these two selection criteria during the innovation period.

Man or Woman?

The rationale for the sex criterion was the management's belief that men were better suited to the requirements of the automatic roles, and that they would also form a more stable and reliable labour force than women.

In Chapter Six we have seen that the management perceived the new role as requiring less manipulative ability on the cloth, and as involving more walking around the much larger setts of looms. Both these views of the automatic role influenced their belief that men were its most suitable occupants. Further, we should not divorce these direct features of the role from the indirect ones such as the shift character of automatic weaving. In the next chapter we shall see that this feature also strongly influenced the management's policy on the sex criterion.

Although more than 80 per cent of the automatic weavers in the new section are male, there are some notable exceptions to this selection rule. When the innovation was mooted, there were already two women weavers on the small group of 48 automatic looms. They formed, in fact, half the nucleus of experienced automatic weavers which played such a vital role in the innovation. Both these women left the company in January 1955, and for the next two months the maleness of the new section was complete. One of the women was probably the best automatic weaver present in the section during the course of our study. This judgement is based on the criteria which have been already described in the last chapter. She, more nearly than the others, fulfilled the role requirements as they are set out, and this was generally agreed by the members of the management group and even by the operatives themselves.

The second exception came in mid March 1955, when two women from the specialist non-automatic weaving section joined the section as trainees. As described in Chapter Five, the management were reluctant to take such a step, but some reshuffling was necessary and no other solution was available if production was to be maintained. In a few weeks these two weavers were responsible for full setts of looms, and at least in terms of quantity of output, they were among the 'better' weavers in the section.

The consequences of this criterion have been felt both within and outside the automatic section. Many women weavers in the non-automatic departments expressed opinions in defence of the traditional sex of a weaver. We were often told about the inability of the men on the automatic looms to mend 'smashes' and to weave 'pick-at-will'. Both these job elements were part of the role of the non-automatic, but not of an automatic weaver. Such statements were also made at meetings of the Works Council, as: 'I'd like to see some of the men doing the things we do.'

This method of defending their own status by belittling some features of the newly exalted men is a good example of the confusion over roles that constantly recurred during the study. When a factory is undergoing such rapid change there are always many people who are using inappropriate frames of reference for the appraisal of various features of the change. Failure to recognize the emergence of a new weaver's role, together with the change in the traditional sex of the operatives, is the basis of these attitudes and opinions.

As further support in their attempts to withstand the cultural change caused by this selection criterion, women weavers quoted the invariable success of those of their number in the section. That this success may have been due to specific abilities and motivation rather than the general factor of their sex was usually ignored by them. This was in line with the general acceptance of the non-specific selection in the life of the mill.

The men in the section, on the other hand, retaliated by emphasizing some of the prominent difficulties experienced by women occupying the role.

> 'They don't manage without help. They wouldn't be able to run 24 completely alone.'
>
> 'They aren't just anyone – they're specially picked women weavers.'

'Women aren't so mechanical either and they will persist with a loom fault which men would recognize easily and perhaps even fix it.'

Some interesting comments by some of the men weavers bear on the hypothesis we have developed about their increasing cohesiveness. These men perceived the presence of the few women in the section as a deliberate attempt by the management to lessen the solidarity of their group. There is no evidence to support the truth of this belief, and the choice of these women by the management appears to have been governed by expediency alone. Nevertheless, it is easy to see how such beliefs could arise from members of a group which is becoming more cohesive and looking for uniformity.

The traditional occupancy by women of the weaving roles had its effect on the general frame of reference held by most of the personnel of the mill. As has been illustrated in the previous chapter, only the top managers in the management group had frames of reference adapted to automatic working. This sex element of the traditional or non-automatic frame of reference often came out when the various foremen were assessing the automatic weavers. 'Most of the automatic weavers are men and they don't make such good weavers as women.' Oakroyd and Laycock, on the contrary, often referred to the experienced male weavers in the section as 'very good weavers'. Moreover, this preference was shown again, in the form of a 'scapegoating' or defence mechanism, during the 'hooter incident' described in Chapter Four. Then they translated the immediate issue into one of 'the unsatisfactory nature of women weavers as a stable labour force'.

A further effect of the new sex criterion in the automatic section itself was to change the traditional relationship between those occupying the two roles of weaver and overlooker. We shall discuss this relationship in the next chapter.

Old or Young?

The management's rationale for the age criterion was twofold.

First, young operatives were believed to be more adaptable and more easily trained. Second, once they had been trained, the company could look forward to a longer period of service from them. The validity of both these beliefs can now be seriously challenged, both in general and in the light of the present study. However, given the experience of management at the time this criterion was formulated, it was certainly a reasonable working hypothesis. This study has afforded us a valuable opportunity to observe a real social system subjecting such hypotheses to the test of experience in its ongoing everyday life.

The first effect of the age criterion was a stiffening of attitude among the older operatives outside the automatic sections to any changes or possible changes in the pattern of the mill. This was particularly clear in their attitude to shift work.

There was a general air of apprehension even among groups that were to be virtually unaffected by the change. Interviews with these operatives have indicated that this apprehension arose because of the changes that were happening elsewhere in the mill. It will be remembered that the management made a positive point of informing the operatives who were to be involved in any changes. It appears that it is also important in an interdependent social system, such as this factory, to inform unaffected groups that they will be unaffected. The importance of such negative information became clear to the managers during the innovation, when they were confronted with the task of dispelling a series of false rumours that had sprung up in various sections of the mill. Our view of the factory as a dynamic and interdependent social system undergoing certain changes would lead us to expect such reactions in materially unaffected parts of the system.

Within the section itself, the age criterion produced a consequent effect very soon after the innovation began. This effect stemmed from the perception by the operatives, in outmoded, non-automatic frames of reference, of their new role and its new selection criterion. The following comments from both overlookers and weavers in the section indicate the result of this perception.

Automatic overlookers:

> 'I think automatics are a dead-end job for men. There's so much pressure on you and it's the speed that counts. For us, it will be a case of whether we can slow them [the management] down to our pace.'
>
> 'I often say now that I hope I won't still be on these when I'm fifty.'

Automatic weavers:

> 'I hope I'm not weaving at fifty. What we're afraid of, though, is that they'll find us jobs with a broom and £5 per week. There are some who have thought that and as soon as they found another job they left. I'm not saying you couldn't get a fair efficiency at sixty, but how many people could? If there are younger people as good, or better, they [the company] won't keep you on.'
>
> 'I just don't know what happens. I can manage now but I'm not going to wait till I can't manage. The company should have some policy about it. What are they going to do?'

This general apprehension about the long-term implications of the new roles was particularly marked among the weavers. No such uncertainty had existed in the mill prior to the innovation, and it appears to be a direct inference from the age criterion. The weavers' strong reaction is explained again by the nature of their group. In most other ways the security of this group had increased because of the innovation. Their perception of this criterion as meaning that the new role was only for young men detracted from this increasing security, and so their reaction was strong. Further, since the weavers were now ranked second to the overlookers in wages, the criterion was interpreted as a threat to their ultimate financial security. Economic factors involve the total life-space of an individual and thus a threat in this area will be seen as a threat to the whole way of life, and reaction to it will be appropriately strong. Six of the overlookers expressed similar opinions, although much less strongly. The security of the overlookers' group in the mill and in the industry, and the ever-present possibility of promotion for its members may help to explain this. However, any expression of such opinions by one of this superior status group reinforced the weavers' interpretation of the new criterion. 'I think age will tell. I don't encourage people to leave but I never discourage anyone who has ideas about another job.'

Among management there was again a difference of opinion. Oakroyd and Laycock were confident that the weavers and overlookers would be able to carry on with increasing age, but others of the management group were not so sure.

The average age of the weavers in the automatic section in 1955 was 28 (age range 16–41) and that of the overlookers was 31 (23–42). Debenham Mill provided no solution to the uncertainty about a limiting age, for a similar policy had been adopted at that mill in 1948–50. This resulted in 1955 in average ages of 31 (21–45) and 40 (24–50) for the weavers and overlookers respectively. However, of interest with respect to both criteria is the fact that the oldest weaver at that mill was a woman, and she was generally assessed by the management there as one of the best weavers.

Some further information that bears on this point was obtained during another study in a mill in Lancashire, which had had its automatic looms for a longer period than either of the present ones. The weavers were all women and averaged 40·5 years (age range 21–60) while the overlookers averaged 46 years (25–66). No uncertainty about this age question or about the ability of women as automatic weavers was detected in this mill.

A TRAINING PROGRAMME FOR THE WEAVERS

The weavers entering the automatic section were essentially of two types. There were those who had been experienced non-automatic weavers, and those, recently recruited, who came straight to the section from the training school. Since the training school at Radbourne had no automatic looms, both these groups came to the automatic section with a similar background as far as machine type was concerned. The Debenham school had had one automatic loom which made some training for overlookers possible. The trainer at Radbourne complained repeatedly that there was not such a loom in her school, but the management did not believe it was necessary. Such a difference of opinion between the production management and the trainers also existed on the question of the cloth from the training looms. The management rationale was that training was not separate from production, which meant that the looms in the school were to be operated as production looms. This was particularly disliked by the trainers: 'We just can't let them go ahead and practice as they should on some of the cloths, because we mustn't get too many faults.'

Another consequence of this rationale was the fact that training in the school was not thought of as providing complete training for production. It was always regarded as preliminary to an ultimate training with an experienced weaver on production looms.

This rationale was opposed in principle by the trainers and by Manning, the personnel manager, but these people had no executive authority, and on this point they were quite isolated from the rest of the company. This frustration was typified by the remark: 'At some stages in this company, it seems to be fashionable to have a training school. Whether it is any use or not is not important.'

The important overlookers' union supported the management rationale in their terms for apprentices, who were trained with a full overlooker on a sett of production looms. As will be seen later, the rationale was also widely accepted among the weavers in the mill and by the weaving foremen.

> 'Weaving is like pneumonia. You don't learn it, you catch it.'
>
> 'I think you can learn much more by yourself in the shed, than by someone telling you.'

Despite the long history of training schools for weavers, they remained strictly limited both in function and in acceptance by the company.

This attitude to weaving training was maintained in the programme for the weavers entering the automatic section. Both types of recruit were to spend a period of time as a trainee weaver with an experienced automatic weaver, before taking charge of a sett of new looms. Initially it was planned that this time should be three months for the experienced non-automatic weavers, and an indefinite period depending on their progress for those straight from the training school.

In fact, owing to the delays and other emergencies described in Chapter Five, the actual times spent in this way by the trainee weavers up to July 1955 are as shown in *Table 8*. The figures in brackets indicate that this time was spent with another 'new' weaver in the section rather than with one of the original four automatic weavers.

TABLE 8 TIME SPENT AS TRAINEE WEAVERS

Type of Weaver	*Time as Automatic Trainees (Weeks)*
Non-automatic	18; 18; 2+ (16); 3+ (16); 10; (1); 2; (16)
Training school	(2); 10; 12; 1

The management believed that the training problems for the weavers were very small, despite their numerous comments, outlined in the last chapter, about the changes in role and function of the weavers. However, it will be remembered that they perceived the main role change to be associated with the multi-loom character of the automatic setts. The traditional training unit was the loom, and such sett-oriented features as supervision and systematic coverage were hence not regarded as part of the training programme. Apart from some of the automatic weavers themselves, this divorce of the sett features of the role from training was characteristic of the view of all members of the mill. The weavers expected and were expected by others to develop such features for themselves.

The training programme in the section immediately introduced a new relationship in the social system, namely that between the experienced weaver and the trainee weaver. With one notable exception, this relationship took the following pattern. The experienced weaver spent a day, or at most several days, showing the new weaver how to stop and start the automatic loom and the other major features of its mechanism. Thereafter, he assigned his trainee to full responsibility for a sub-sett (usually half the looms) while he operated the remainder. Such independent working continued until the trainee was removed from the sett by the management. Sometimes the experienced weaver would help the trainee if special difficulties arose, or if he considered the trainee was not achieving sufficient efficiency from the sub-sett. The exception was the relationship between the most experienced woman weaver and one of her trainees. She alone maintained effective control of her full sett, and allowed the trainee to work with her on isolated looms that stopped.

In only one instance did the management have to separate a pair of weavers because they were unable to make a satisfactory personal relationship. The effectively independent nature of the usual working relation between trainer and trainee, as just described, minimized these potential relationship problems.

Nevertheless, difficulties of another kind did arise out of the relationship. As soon as looms earning began in December 1954, the trainees expressed discontent to the management.

> 'I want a sett so that I can earn more money. Now I'm working for Joe.'
>
> 'No matter how hard I work, I can't earn any more.'

The experienced weavers retaliated by criticizing the trainees, although they all gained by the relation as it was practised. 'I think you should get paid to teach, but they [the management] don't see it that way. That's why some weavers don't like learners.'

It will be remembered from Chapter Five that the fixed wage rates of the trainees were readjusted at this time, but the tensions continued until the new looms arrived, and the trainees were able to have setts under bonus conditions themselves.

ASSESSING THE TRAINING PROGRAMME

During this training period, all the trainee weavers, except two, expressed the opinion that they ceased to learn long before they gained full setts. This was even the case for those who were trainees for only two weeks. One exception was an adult weaver who appreciated the easier working conditions of training and was quite satisfied with his average wage. He was one of the isolates among the weavers in the section, to whom we have already referred, and his atypical opinion is further evidence for so classifying him. The other exception was a juvenile weaver who was very apprehensive of facing full responsibility, and 'didn't mind how long [he] waited for a sett'.

Towards the end of the study, when all the weavers had been on full setts for at least six months, they were again asked to assess their training period. Thirteen of the fourteen weavers reaffirmed the belief that the period was unnecessarily long, while the isolate weaver retained the contrary opinion.

With such varying training experience, no minimum time for training was agreed upon by the whole group. Those whose training had continued for long periods while the looms were delayed tended to protect their own prestige by saying that longer periods were more necessary than the very short periods experienced by others. 'The only reason they got away with it was the easy quality that was in the looms when they started.'

This general opinion among the weavers about their own training is thus seen to conform to the attitude which prevailed about it throughout the mill. Their experience of the new role did lead them in general to perceive sett-centred features, in particular, the need for a systematic method of working. Nevertheless, they did not associate these features with training. The training problems were all related to the operation of a single automatic loom, for which they felt their training was more than adequate.

There were, however, as mentioned earlier, indications from a few of the weavers, that the change in their frames of reference to an automatic orientation also led them to view training in a new way. One did criticize the whole training situation and two others expressed the belief that systematic working could be taught to some extent.

> 'This initial training period is so artificial with three overlookers and eight weavers on 48 looms. It's wrong to think that there is just twice as much work in running 16 looms as there is on 8.'
>
> 'I think it could be taught, but nobody has ever mentioned it.'
>
> 'No one ever told me about a system. I worked it out but it depends on the conditions and you have to alter your system to fit. . . . Yes, I would tell anyone I had to train.'

It would be interesting to observe the continuing impact of experience in the automatic roles on the established attitude to training in the mill.

TRAINING PROGRAMME FOR THE OVERLOOKERS

When we turn to consider the training of the automatic overlookers, we find a much more elaborate programme, stemming from the management belief that here the training problems were very considerable.

> 'Automatic overlooking is very different from non-automatic, due to the more complicated character of the automatic looms and to their precision nature. . . . The overlooking is so important that the whole success of the change depends on them.'
>
> 'These overlookers are the key people and for them things are very different. . . . For an old-type overlooker, the problem is largely one of forgetting what he already knows and being prepared to start afresh with the new material. The old loom overlooking was very much learnt and picked up on the loom. Now there is a manual, and all the information is there.'

'Overlookers will be very difficult to teach, for there is no longer individual character for each loom, but critical settings of precision machines.'

The five overlookers who entered the section had all had experience with non-automatic looms but otherwise their experience was of very varied content. From the management comments, it is evident that this previous experience was regarded as something of a mixed blessing.

The training programme was as follows:

1. The new overlookers were attached as trainees to the three experienced overlookers in the section until the new looms began to arrive. As in the case of the weavers, this was in accord with the traditional method of training apprentice overlookers.
2. Kaye, a former automatic overlooker, now attached to the company head office, conducted a training class for between six and eight afternoon sessions, working from the manual supplied with the automatic looms.
3. An instruction film on the loom was shown by Kaye.
4. The new overlookers assisted in the assemblage of the new looms, thereby gaining experience of the looms under non-production conditions.

Owing to the unforeseen delays, the overlookers spent varying times on the first and major part of this programme. One had been attached as an apprentice to the experienced automatic overlookers in this way since early 1954, and the others had 18, 15, 8, and 10 weeks respectively.

We have described in Chapter Five the early innovation period from January to June 1955 when the looms were arriving, and there we saw that four other features were added to this training programme.

5. Kaye worked in the section for six weeks assisting the new overlookers on their setts of looms.
6. One of the overlookers went to Debenham Mill for ten weeks training, enabling

7. Knight, an experienced overlooker from that mill, to take charge of one of the new setts at Radbourne, and to teach the new overlookers during overtime working and when he was not fully occupied on his sett.
8. The overlookers worked overtime for a short period 'to enable shift opposites to be together on their looms and to learn to work in with each other'.

The elaborate character of this programme contrasts with that for the weavers. However, it was still entirely consistent with management's limited perception of the sort of role changes which would be involved for the overlookers. Apart from Laycock, the management perceived the changes in loom-centred terms only. These loom-centred changes were very considerable and all these lay within the scope of training as it was accepted by the mill. Knight, like Laycock, also recognized sett-centred features in the new role. However, apart from his contribution to the training programme, all other parts were aimed directly at these loom-centred features.

As well as expressing concern about the magnitude of the training problems for the overlookers, several members of the management were uncertain of the overlookers' reaction to any training programme. They were very relieved when the programme was readily received by the overlookers. 'To get them to agree to a training course has been quite an achievement. They already regard themselves as full men.'

The implications of this misjudgement of the overlookers' response have a considerable bearing on the relations between groups in hierarchial social systems like factories, companies, and offices: a point that will be taken up in the next chapter.

The acceptance of the training programme by the overlookers appears to be quite consistent with the analysis of their group as it has been developed so far. It also represents a very successful attempt by the management at providing training for technological change.

Although the management drew up the training programme, they played little direct part in its application. Parts 1, 4, and 8 were entirely consistent with the training norms accepted by the overlookers – namely, training is internal to the group. Parts 2, 3, 5, and 7 did involve two people external to the Radbourne group, but Kaye and Knight, on the other hand were not members of any other group at Radbourne, most particularly not of the management group. Kaye had been an overlooker at Debenham until recently, and during the innovation he was in a company position which was not identified with the local production or executive management. Knight was currently an overlooker at Debenham Mill, and it is of interest to note, that he did not agree to Laycock's request to transfer to Radbourne Mill, until he had gained the approval of the overlookers at that mill. The two 'outsiders' did not, therefore, threaten the cohesiveness of the group, nor were its traditional norms disturbed, as they would have been by active participation in the training by Laycock or some other member of the management group.

Within the group, the programme did involve the overlookers in new personal relationships similar to those described for the weavers. This training relationship with the experienced overlookers was very successfully carried out. The relationship remained a typical training one throughout the periods given above and did not develop those aspects of independence observed in the case of the weavers. The many new features of the new looms as machines accounted in part for the fact that the new overlookers did not become independent. They simply could not have coped alone with even a sub-sett of the looms without seriously lowering production. However, the personalities of the experienced overlookers and their ready acceptance of this training of their fellow group members were very largely responsible for the success. Good relationships were easily achieved and maintained.

'It made all the difference to me. He really had them taped. Then he was my type – I could work with him.'

'Young and Irwin are good chaps and they are very helpful.'

Finally, there was not the same wage differential between the experienced men and the trainees as for the weavers, so this did not provide a focus for dissatisfaction with the relationship had any existed.

Other new personal relationships appeared between the trainee overlookers and the weavers. The marked disparity in the difference between the overlookers' abilities during this training period was a source of irritation to the weavers, particularly those on bonus earning. The difficulties that arose in these relationships are indicated by the enthusiasm with which the new overlookers greeted the fourth part of their training programme: 'Putting the looms together enables you to make mistakes, and they don't show up. If you get into trouble, you can wait and get the help of the experienced overlookers. It's nice not to have any weavers around.'

The eighth part of the training programme was aimed at yet another type of new relationship for the overlookers, and this is also taken up in detail in the next chapter.

ASSESSMENT OF PROGRAMME

Both during and after the training period, the overlookers and the management continued to view their problems in training and in normal production in loom-centred terms. From the beginning, the trainee overlookers were very conscious of the magnitude of their changed role. Throughout their training they were appreciative of the value of all parts of the programme. In contrast to the weavers, there was always a feeling that the training could have been longer and more extensive. At the end of the study, the overlookers all agreed that they had still more to learn, especially concerning different qualities of cloth in the automatic looms – again a loom-centred feature. Under the pressure of production, and because of the very importance of the overlookers to the success of the innovation, the management were often impatient with the progress of the training. The additions to the programme bear this out; but at other times they expressed very great satisfaction at the overlookers' achievements. Certainly, the training of the overlookers in such a short time was a major factor in the success of the whole innovation.

REFERENCE

Work not specifically referred to in the text

KING, S. D. M. (1960). *Vocational training in view of technological change.* Paris: European Productivity Agency, Project 418.

EIGHT

Changing Relationships

At the beginning of this book we said that any technical change in a factory can be regarded as so many changes in the human relationships in the social system defined by the factory.

Because of the change, individuals and groups find themselves in new roles. In order to fulfil these new roles, the individuals and groups find themselves in relationship with new individuals and new groups. Furthermore, many individuals and groups whom they continue to meet, now stand to them in different sorts of relationship from those existing formerly.

It is this view of the innovation of the automatic looms that we wish to take up in detail in this chapter. In Chapter Five, we have already discussed some of the changing relationships within the management group, and we now consider changing relationships among the operatives.

We have seen that the two main groups of operatives in the automatic section changed in their functions and in their character as groups. The automatic weavers, in particular, became a more cohesive group and they also increased their status in a number of ways. The automatic overlookers, on the other hand, remained essentially part of the wider and already highly cohesive group of all the overlookers in the mill.

Our analysis of the changing patterns of human relationships is based largely upon this overall change in the character of these two main groups. Factors arising from the groups to which individuals belong have a large influence on the relationships that are established between them. However, it is necessary to know what relationships existed prior to the innovation if we are to appreciate the changes that occurred.

NON-AUTOMATIC WEAVING

First, we will consider the traditional pattern for the weaving department under non-automatic conditions.

Each weaver would have charge of from 4 to 8 looms and an overlooker would have a sett of 40 to 50 looms. This meant that each overlooker came into a working relationship with anything from 6 to 12 weavers. It also meant that each weaver, while in attendance on her sett of looms, was in comparatively close proximity to several other weavers. Certainly, with the aid of lip-reading, which was highly developed among the operatives, they were able to communicate with each other from time to time without leaving their looms. In the post-war years other types of personnel also appeared in the weaving shed. Various parts of the weaver's role became the specialist functions of a number of ancillary operatives. Weft and packet carriers, cleaners, helpers, and finally oilers were all established as roles in the operative structure, and the role of fitter also appeared as an ancillary to the overlooker. The respective work-tasks of all these operatives brought them into some relationship with the weaver and the overlooker of any sett of looms. *Table 9* gives the average fraction of the working time of each of these types of operative that would be spent with one particular weaver – or at least spent in the geographical area of one weaver's sett of looms.

The most striking feature of this non-automatic arrangement is the fact that only the relationship between a weaver and her overlooker provides for any appreciable amount of overlap, when they are actually involved together as a working combination. This overlap was even less before the existence of the ancillary roles when weavers had only 2 looms each.

There was a relation between each weaver and the surrounding weavers, but this was not an official one as far as their respective work-tasks were concerned. However, in practice, non-automatic

TABLE 9 NON-AUTOMATIC WEAVING OPERATIVES' TIME RATIO AND LOOM COVERAGE

Role	*No. of Looms per Operative*	*Average Overlap of Working Time per Single Weaver*	
		Fraction of Total Working Time	*Fraction in Hours per Week*
Weaver	4	1·0	45
Overlooker	40	0·1	4
Fitter	200	0·02	0·9
Cleaner	200	0·02	0·9
Oiler	200	0·02	0·9
Weft and Packet Carrier	150	0·025	1·13
Helper	100	0·04	1·8
Weaving Supervisor	300	0·013	0·56

weavers did help each other on occasions, and this mutual working together reinforced the communication contact that was possible for these operatives because of the proximity of their working areas. Such 'help' was, however, largely given or withheld by the free choice of the weaver, and was not an obligatory part of his role.

Besides these relationships which are a direct consequence of the work-task or of their position in the shed, the individuals in the factory participated in a number of other personal relationships. Each individual had certain other individuals with whom he shared a social or informal relationship. These relationships took the form of face-to-face communication, unofficially during working time and officially during periods of leisure such as tea breaks, visits to the toilets, and meal times. Often these social relationships were extended beyond the confines of the factory and the hours of work. This type of relationship, in particular, appears to play a vital part in the life of social systems such as our factory, and more particularly in the life of the role groups. For example, it was largely through such relationships that the weavers developed their cohesiveness and consequent favourable attitude to the innovation. Much more study of the relationships in these role groups could well prove rewarding. Although officially recognized neither by the terms of the work task nor by management in general, the relationships are in other ways quite formally written into the social structure of the factory.

As in a number of other cases (for example, Homans's findings among groups of office workers), it appears that informal relationships, as well as formal or work relationships, are determined by the existing social system. That is, they lie within the role groups as defined by the formal social structure. In other words, weavers develop these social relationships with other weavers, foremen with other foremen, overlookers with overlookers, and so on. The more closely defined and cohesive the role group, the more generally does this pattern hold.

Conversely, the more cohesive the group to which an individual belongs, the more difficult is it for him to develop social relationships with members of other groups. To do so tends to make an individual suspect among the remaining members of his own group. This hypothesis also finds support from the study by Homans mentioned above. He found patterns of sociometric choices (i.e. relationships based on personal liking) and social interaction which usually lay within role groups. The general validity of this hypothesis and its implications require further investigation. Moreover, the pattern of these relationships was formally fostered by the structure and culture of the mill. The structural divisions of top management, foremen, and operatives were maintained during the meal breaks by the existence of separate eating facilities for these groups. In the case of the operatives, there was a further division by sex and this also effectively divided one set of role groups from the others, since most roles were held by one sex only.

Again, the operatives were themselves organized in unions which followed the same boundaries as the role groups. Thus the role an individual held determined the union to which he was eligible to belong. Related to this, for it is part of the *raison d'etre* of unions, is the fact that payment was to the individual according to some arranged terms which applied to all occupants of his particular role. Finally, in the Works Council, a cooperative venture of management and operatives, some of the representation was again in terms of the role groups and levels which appeared formally in the social structure. The remainder of the representation in the Council stemmed from another formal element in the structure – 'the department'. Thus, one councillor represented the warping department, which included individuals occupying several types of role, another represented the engineers, and so on.

WORK GROUPS AND THEIR NON-EXISTENCE

The important thing to note about all these aspects of the social structure and culture of the mill is the absence of any clear definition or formal expression of the group of persons whose interrelations were dictated by the nature of the work task. Such a group we will call a 'work group' as distinct from the role group which has been our main subject for discussion so far.

A work group for the task of weaving under the non-automatic conditions we have described above would consist of a weaver and one of each of whichever of the other role types (see *Table 9*) existed. Added to these would be one or more weavers who lent unofficial assistance, and the weaving foreman and other management members who directly contacted the weaver in the fulfilment of his role. Thus, this group would essentially contain members whose roles were different.

This group was not formally recognized in the structure or in the culture of the mill and even in the work task the extremely varied overlap of its members in both time and space made its definition obscure. From what has been said already, it is clear that the formal structure of the mill was not in terms of these work groups, but rather emphasized role and departmental groups. Human relationships tended to follow this structure as did a number of other features in the life of the mill, such as union membership, representation on the Works Council, and the geographical organization of production.

AUTOMATIC WEAVING

In the light of this distinction between role and work groups, we will now turn to the pattern of the automatic weaving section. A weaver now had charge of between 16 and 32 looms, which were operated on a double-shift system. The overlookers now had 40 looms. The pattern for all the operatives in the section is given in *Table 10*.

By comparison with *Table 9*, it is seen that all the overlap times have increased considerably. The overlap with the overlooker approaches totality, and not more than three weavers are now in relationship with any particular overlooker. In the present case, battery-fillers were not used, and in factories where they were the weavers' setts would be even larger, further increasing the overlaps. The other large overlap occurs with the helpers and spare weavers, who were all women and usually former non-automatic weavers. The automatic weavers, as we have seen, were nearly all men.

Of the individual operatives, who are represented in the table, only the weaver, overlooker, fitter, and cleaner spent their whole time in the automatic section. The others operated in work domains which included the whole of the main weaving shed with its three different weaving sections.

Because double-shift conditions applied to the automatic section, each sett of looms was served by an identical group of operatives with similar overlaps on the other shift. These shift conditions brought these two groups of operatives who shared the

TABLE 10 AUTOMATIC WEAVING OPERATIVES' TIME RATIO AND LOOM COVERAGE

Role	*No. of Looms per Operative*	*Average Overlap of Working Time per Single Weaver*	
		Fraction of Total Working Time	*Fraction in Hours per Week*
Weaver	24	1·0	37·5
Overlooker	40	0·6	22·5
Knotting mechanic	200	0·12	4·5
Fitter	160	0·15	5·6
Cleaner	160	0·15	5·6
Oiler	200	0·12	4·5
Weft Carrier	160	0·15	5·6
Spare Weavers and Helpers	80	0·3	11·2
(Battery-filler	80	0·3	11·2)
Shift Manager[1]	160	0·15	5·6

[1] In so far as the shift manager was the direct weaving supervisor; however, his figures would in practice be somewhat reduced because the shift managers did have wider responsibilities as well which took them out of the weaving shed.

same machinery into a work relationship with each other. We will discuss these relations between shift opposites later in the chapter.

Another difference between the automatic and the non-automatic patterns lies in the weaver–weaver relations. The introduction of work-study methods had preceded the innovation; nevertheless, it was the prospect of the automatic looms that largely prompted the company to introduce these methods of assessing work-loads. One consequence of work-study was to even up the weavers' loads. With this general levelling, the weavers found they were more constantly involved with their setts than previously, although never as highly as on odd occasions previously. Accordingly, they found that they neither needed as much assistance from other weavers as previously, nor were as free from their own setts to give aid to other weavers in difficulty.

Weaver–weaver cooperation and mutual help became a much less prominent feature of the working pattern with a consequent loss in comradeship between people with the same role: 'It's a funny thing, but in the past a weaver would often help another weaver who had a lot to do on a bad smash, but now you don't get much of that sort of thing. Everybody is more independent.'

This found its fullest expression with the multi-loom setts of automatic weaving. Not only were the loads still more even, but the work task was now quite continuous, allowing no time for such weaver–weaver working relations. Finally, with 24 or more looms in his sett, the weaver found himself in a working domain which isolated him from the next weaver by distances of between ten and twenty yards.

These factors, together with the increased noise level from the automatic looms, made face-to-face communication between weavers while on their setts virtually impossible. To interact in this way, one or both of the weavers had to leave their sett of looms unattended.

We thus see that the nature of the weaver's new role prevents either a social or working relationship with other weavers. Nevertheless, we have seen that this type of relationship remains the one which is supported and finds natural expression in the structure and culture. The automatic weaver who fulfils the requirements of his role is now left with only the working relationships with individuals occupying roles other than his own – in other words relations with members of his work group. The automatic work group included with the weaver, one of each of the other role types in *Table 10*, the shift manager, and to a very limited extent the assistant manager and the manager of the mill. Despite the increased overlap, the group received no more formal expression in the structure and the culture than did the non-automatic work group. We now have a basis for understanding the apparent paradox that a favourable attitude towards the innovation did not always correlate with the weaver's fulfilment of his new function. The weavers who had favourable attitudes to the innovation were those who participated in the growing cohesion of their role group. The new frames of reference, which grew out of this group participation, led to their favourable attitudes. But the very growth of this new group with its particular character meant that the members tended to interact more. This weaver–weaver interaction under automatic conditions is achieved as we have seen at the expense of the fulfilment of the weaver's automatic task. On the other hand, the comparative isolates among the weavers held most strongly to old frames of reference and unfavourable attitudes. Nevertheless, such non-participation often meant that they were in attendance on their looms more continuously than the other weavers, and so they fulfilled their task better.

A CASE OF WORK GROUPS FINDING FORMAL EXPRESSION

Rice (1958) has described certain reorganizations in the social structure of an Indian weaving mill. This situation is of particular relevance to the present analysis because the reorganization was based on the work group as we have considered it above.

The Indian shed initially had a structure based on role groups, which was similar to the one we have described for Radbourne Mill. The group of individuals whose work tasks were interdependent did not find expression in the structure, and because of their different loom setts they had little overlap, which further obscured their interdependence.

The confusion of these work relationships was further maintained by the management and the operatives alike, in the titles of the various work roles and the status accorded to them. These titles, as in the present case, were carried over from an earlier pattern of working, although the actual functions were markedly quite different.

An examination of the geography of the looms in the shed, and an analysis of the actual functions performed by the occupants of the different roles led to a suggested reorganization. The different functions were classed as long or short loom-stoppage types depending on their time of direct contact with a single loom. By making some of these functions interchangeable, work groups of seven operatives were found to be sufficient to carry out all the operations in a domain of 64 looms. This domain was the natural geographical unit in the mill. Previously, the diffuse work group had eleven members all occupying different roles as designated in the social structure. The operatives in the new work groups were now to be ranked in only three status grades – A, B, and C – which were deliberately not associated with the names of the old roles.

As soon as this suggestion of a group of workers for a group of looms was put to the operatives' representatives, it was spontaneously accepted. The operatives quickly organized their own groups on the new pattern. These groups then began working for an experimental period.

One result of the experiment after five months was that internally led groups had been created following their formal definition by the reorganization. The old occupational titles were abandoned and reference was only to the three grades. The number of individuals reporting to the supervisors was considerably decreased, strengthening control and allowing the higher management to withdraw their governing influence from the shed. A new level of performance was achieved in which efficiency was higher and damage lower than before the reorganization.

These results, and the very fact that such a reorganization was achieved, must be considered against the background in which this Indian mill was set. The economic, cultural, and industrial conditions of Britain are so different that it is not possible to predict whether such a reorganization would have been possible in the mill at Radbourne. Nevertheless, the innovation there did mean that a similar diffuse differentiation of function had arisen, together with the disappearance of formerly important relationships which were still defined by the social structure. No official re-integration of the work tasks had appeared so that the formation of work groups in keeping with the new work relationships was impeded.

We shall now consider some evidence for the emergence, among the automatic operatives and among management, of behaviour patterns and attitudes appropriate to such a re-integration, which occurred despite the obscuring effects of the formal structure and the co-existence of non-automatic patterns.

WEAVER-OVERLOOKER RELATIONS

In Chapter Six we saw that the increased continuity of automatic weaving emphasized the importance of the relationship between a weaver and an overlooker. An important aspect of the weaver's role was to act as an observer and reporter of loom behaviour to the overlookers. Automatic overlookers, in turn, should also undertake part of this supervision of the sett. The overlooker and his weavers have always been interdependent, but under automatic conditions this contact is further increased by the possible sharing of this supervisory function.

Continuity of production, now a priority in the automatic case, depends largely on a smooth relationship between these two individuals. The weavers in the section certainly experienced this and often commented upon it, particularly during the first months with the new overlookers: 'It's absolutely essential. If you can't get on with your overlooker and the other two people, you might as well pack up.'

Likewise, this relationship weighed upon the overlookers. Under non-automatic conditions, an overlooker had relationships with up to twelve weavers, and if one or two of these relationships were difficult personally then satisfactory human relations could be found with some of the other weavers. Now in the automatic section, the overlooker had relations with only two weavers – each one occupying about half his working time. It is clear that the breakdown of one weaver–overlooker relationship now seriously jeopardized production and human satisfaction.

During 1955, Laycock transferred one weaver because of such a breakdown in her relationship with the overlooker. Laycock himself said: 'When I was an overlooker, I had nine weavers to work with, and now they only have two. But each relationship is now much more important. You now just have to do something about it. You just can't leave it alone.'

In Chapter Seven we referred to the effects that the sex criterion of selection had on this relationship. Traditionally, overlookers were men and weavers were women; and the accepted relationship between them reflected these sex differences. The respective roles had remained sharply distinct and their status reflected male–female attitudes in the general culture.

With the introduction of male weavers, particularly in the automatic section, much of the basis of this traditional relationship and the status of the roles was removed. The weavers now tended to perceive their work relationships with the overlookers in terms of the equality that existed outside the factory for them all as men. They no longer readily accepted the wage differential between the two roles, or the fact that overlookers could look to promotion, whereas weavers could not. Supporting this perception of equality, was the fact that a number of the weavers had the mechanical ability and experience to attempt to carry out some operations on the looms traditionally left to the overlookers. The overlookers resented this threat to their status and prestige on the one hand; as one commented: 'Well, I always prefer women weavers. They don't interfere with the looms. Some men are absolute menaces. They think they know – and make alterations. We don't know where we are.'

But some, on the other hand, found new and favourable possibilities in the relationship with male weavers.

The continuing emergence of this relationship between the men occupying these automatic roles will depend on the interplay of the opposing factors which have just been outlined. At Debenham Mill the sex criterion has not been adhered to over the years, and the presence of numerous women has maintained the traditional status differential, although the functional roles have changed in the same direction.

WEAVERS AND ANCILLARIES

The increased differentiation of the weaving function in the automatic section raises the question of the weaver's relationship with the ancillary operatives. In fact, the weaver was no longer an independent operative, but a dependent member of the work group. As we have seen, there was only a small overlap in time between these individuals and a lack of definition of this group. Thus the relationship was not recognized as important either by operatives or by management at an interpersonal level. The management were very conscious of it at the interdepartmental level, as we have seen in Chapter Five. There we saw that it was important that the weft, warp, and weaving departments should work in close relationship with each other. This would seem to imply that some attention should be given to the individual expression of this relationship in the weaving shed, namely, the group consisting of the weft carrier, knotters, helpers, weaver, and overlooker.

INTER-SHIFT RELATIONS

As stated earlier, during one day each automatic loom was tended by two parallel shift groups of operatives working in the roles listed in *Table 10*.

At both Radbourne and Debenham Mills, only a double-shift system was operating up to the completion of this study. Plans were being drawn up to introduce some three-shift operation of the automatic looms, but their execution was contingent on the general condition of the industry and on many factors relating to markets, union relations, staffing, etc. Nevertheless, three-shift systems do operate in some weaving mills in Britain, so a logical extension of the present case would be the sharing of each loom by three or four separate groups of operatives.

Any such shift system brings these groups into relationship with each other. This relationship is entirely defined by the working conditions and differs in a number of ways from the relations we have already considered.

The individuals are brought into these relationships because of their common involvement with the same sett of machines. In theory, they would not overlap at all in time, and even in practice it was very rare for overlap to exceed twenty minutes per day or one-twentieth of the working time. The members of one group carry on the operation of the machines from the point where their opposite numbers in the other group finish. This means that the state of the machines, or the position of the work or total production, is always new and unknown for the operatives beginning any one shift. Such a situation contrasts strongly with that in the non-automatic departments on day-work. Here, each operative is the only one of his type associated with a particular machine, and all the production from it can be related to his fulfilment of his particular role. When he commences his day's work, the machines and the production are in exactly the condition in which he left them on the previous day.

Thus a relationship between these shift-opposites is one which finds no parallel in the factory on day-work. The importance of the relation between the individuals concerned depends upon the nature of the work task. If the task is discontinuous and made up of units requiring short times, the production of each shift can be kept fairly distinct. But, in the automatic section, production was continuous; shift-opposites were thus related through a shared output, as well as through their interdependence with other operatives, whose tasks might also be continuous, and through their shared machines. The complexity and the precision of the machine would appear to be of importance. If the machine is either in perfect condition or in a state of non-productive breakdown, then any difficulties with it could again be related to a single shift, as would be the case with a discontinuous work task. However, most machines tended to exhibit conditions on a continuum between these extremes and breakdown could often be temporarily postponed. This tendency is probably related to the complexity of the machine. However, its greater precision appears in some cases to have offset this. A further consequence of increased precision is the need for handling the machine according to standardized methods. That is, individualistic behaviour by a weaver or an overlooker can lead to inefficiency which affects both shifts. The automatic looms operating under shift conditions thus have several characteristics which make the relations between shift-opposites even more important than they would be in non-automatic shift working.

The operatives of the work group are more interdependent than in the non-automatic case, and so the relation between shift-opposites can involve individuals occupying a number of roles.

Finally, the increased size of the weavers' sett of looms makes faulty operation less obvious. Faulty looms are more frequently undetected by the operatives on one shift, so that they confront those on the next upon their arrival. There is also a temptation for such faulty operation to be left uncorrected deliberately by the operatives of one shift. Correction or a major repair would mean loss of production and under the payment system only quantity of production was rewarded. The bad quality which ensued led to verbal reprimands only, and even these may fall on the shift-opposite and not on the responsible operative. In this way, there was always a tendency to postpone the final breakdown, which could mean considerable loss of production to one of the shift groups. The importance of this relationship in the section cannot be over-emphasized and some of its various aspects will now be discussed.

The Effect on Status

The status of a role in the mill was determined by, among other factors, the extent of the total production of a particular product for which its occupant was responsible. The differentiation of the weaver's role and the introduction of ancillaries had reduced this responsibility; but the introduction of shift work still further affected it. A non-automatic weaver was associated with every inch of cloth from his looms and accepted responsibility for it. In the automatic section, the weaver was associated with only some of the cloth from his looms, and he now did not readily accept full responsibility for even this cloth. In practice, he was expected to do so, but both he and the management realized that happenings on the other shift could affect the product of his own. Those operatives who were formerly in the non-automatic departments perceived these changes as a decrease in their status, as was reflected in comments such as the following: 'When I trained on the non-automatic looms, we did everything – including the jacquard harnesses – and every piece of cloth that came from the looms you knew was yours. This is all being destroyed – the pride of work and the craftsmanship is going. You never really know if a fault is yours. There is always the possibility that it may have happened on the other shift. You don't blame the other chap but you don't accept the blame either. You get so you just don't care.'

At least in the case of the weavers, we have seen that other factors outweighed this one and led to an increased status for the role. Nevertheless, in a system where conflicting status factors like this exist, we can expect to find dissatisfaction, and the attitude of the overlookers in particular reflected this lack of alignment.

The Effect on Quality and Quantity

The system of quality inspection and sanction in the mill was based on the individual operative and was unchanged for the day and shift departments. The cloth room foreman summarized the position: 'Quality is always worse from shifts. There is the everlasting source of bad quality and trouble in the fact that two people have the same sett of looms. We try and check each shift but they can always blame the other weaver. It happens a lot. Any quality bonus would have to be on the sett of looms.'

The operatives made considerable use of this uncertainty about the origin of damage to rationalize their own behaviour and to thwart the sanctions of the management.

The quantity of production from the automatic section was largely determined by the state of the looms rather than by the individual performance of the weaver. Accordingly, shifts of automatic looms varied less in their production than did shifts of non-automatic looms. The figures for the efficiencies in the section (cited in Chapter Four) bear this out. However, the relationship between the shift-opposites could still affect the production in a way which was perhaps more important to the operatives themselves than to the management. If the deliberate postponement of the breakdown of a loom is practised by one operative then the production of the other shift may suffer a little. The unobserved faults of the machine should even out for both shifts over a period of time, but over short periods, such as the working day or the pay week, they can have a disproportionate effect similar to that produced by faults deliberately uncorrected.

The weavers and overlookers were aware of the possibility of these deliberate and chance causes; but the actual occurrences were still a great source of irritation. For example:

> 'You have to realize that if there's a mess, it isn't always left deliberately. Maybe your oppo. couldn't help it.'
>
> 'It can be a great source of irritation. My oppo. is all right, but the other day he forgot to fasten the reed on a warp he was gating. I spent two hours the next morning getting into endless trouble with it. When he came in he told me that he'd forgotten, but it was annoying and enough to make you fed up.'

These difficulties over the quantity of production were manifest on a number of occasions and have been referred to in Chapter Four; they were heightened when some weavers were on bonus wages and others on a fixed wage. One weaver commented on this: 'The looms do not run well. The trouble is the other weaver. He is on bonus and doesn't worry about anything as long as the loom is still running. He doesn't straighten selvedges or attend to things.'

Management intervention in such cases was of little avail unless it took the form of the drastic step of breaking up the pairings and moving one or both individuals to other setts of looms. This action did not repair the relation, but rather eliminated the need for it by creating two or more new ones which might or might not operate more smoothly. Most of the operatives accepted these difficulties as problems to be solved within their own group. That is, the relational character of the difficulty was recognized, and resort by one of the opposites to the management was frowned upon. This group attitude of the operatives was illustrated in the case of one of the pairs of weavers.

One of these weavers had been in the section longer than the other and he assumed that he was in a position of seniority, which allowed him to dictate the terms of working. He said: 'I have spoken to her many times and told her that I have the experience and that I know about these looms.'

Any such seniority was unrecognized by the other weaver, and indeed by anyone else in the mill. The first weaver repeatedly accused the other of keeping the looms running when they needed the overlooker, thereby decreasing his own efficiency. Finally, he took his complaint to the two shift managers. They were unwilling to alter the pairings of the weavers and the relationship steadily deteriorated until it was resolved when one of the weavers left the company.

The other weavers deplored the action of reporting the difficulty to the managers; for example: 'People too often find a mess and moan about it, or complain to the foreman. You have to work it out for yourself. It's no good going to the foreman and getting him to see your opposite. His only way of solving it would be to move people around.'

The pairing of the overlookers in the section provides another example of the importance of this relation, and of the difficulty of management action in such situations.

In planning the pairings in the group of automatic overlookers, Oakroyd and Laycock were faced with the fact that the eight overlookers were very uneven in ability and experience of the looms. We have already discussed in Chapter Seven how they made use of these individual differences in the training programme for the overlookers. The pairing plan was based on a rationale of again using the experienced men as fully as possible. Young and Irwin, the two original overlookers, started all the new looms as they arrived. The other overlookers, in the order of their experience in the section, were left in charge of the setts as they became ready. This meant that Downham, the other experienced automatic overlooker, was opposite Pierce, the apprentice from the group of 48 looms. Illingworth, a Lancastrian overlooker of considerable experience, was opposite Evans, who was previously an apprentice in a non-automatic section. Hambly and Robson, non-automatic overlookers of similar experience, took the third sett; and Young and Irwin remained on the fourth sett when it was in operation. Thus the plan meant that two pairs were unbalanced in terms of general overlooking experience, and it should be remembered that such experience counts strongly towards the status and prestige of overlookers, particularly in the non-automatic tradition. Young and Irwin, whose cooperation was so essential to the initial success of the innovation, favoured this plan, since it enabled them to retain their already excellent relationship. The others expressed doubts about it, and continued to criticize the plan and its rationale in the first months of operation of the new looms, as is shown in the following comment: 'It's best to let people sort themselves out. Too often in this factory the management put good people with good, or else a good person with a bad person, hoping they'll both become good. The bad one can't keep up with the good, and the good has a lot of extra work to do which leads to moans and complaints.'

Laycock, in particular, was not unaware of the problems that can arise when such pairings are arbitrarily set up by management. He had been involved in a similar situation at Debenham Mill. Two overlookers had disagreed on the adjustment of the tension stop. This difference of opinion finally led to a serious drop in efficiency from their sett of looms, because each overlooker began his shift by resetting all the looms to his particular adjustment. At length they came to Laycock, the senior foreman at the time, and asked him to decide who was right. They were encouraged by the remainder of the overlookers, who knew that Laycock would be unable to decide, since either adjustment could be used without ill effect to the loom or the cloth. Since the inefficiency had been noticed by the management, the group of overlookers felt that this approach would, in some measure, compensate for the admission, to the management, of bad relations between two of their members. However, Laycock, realizing his dilemma, called all the overlookers together and said he would not decide. Rather he gave them a free hand to sort themselves out in pairs of their own choosing, and to work in that way thereafter. This intra-group approach, which proved so successful at Debenham Mill, was included in Laycock's plan for Radbourne Mill. As soon as all four setts of looms were operating, Laycock called the eight overlookers together and offered them just such a reshuffle. The overlookers replied that no changes were necessary, and Laycock confidently expressed the opinion that there 'will be no more trouble over the overlookers' pairings.'

However, over the ensuing months the tension in the uneven pairs mounted. Complaints about their opposite's behaviour were first made by individual overlookers to other overlookers, and then to their union representatives. Changes in the pairings were demanded, and the union, after discussing the problems, drew up a plan which involved altering three of the pairings and even the shifts of two overlookers. The union then approached the management with this plan, which was readily agreed upon, and the changes then took effect.

The strength of the feelings in this small group of overlookers is apparent from their comments at the time of the reshuffle.

> 'It wasn't fair that the two most experienced overlookers were together, while two of us had only apprentices opposite. You get more work to do and your bonus suffers because of the other shift.'
>
> 'There was no pressure from the management. Now it means that less experienced chaps aren't together. My new oppo. had bad luck in his apprenticeship. Now he's a full man, but he's keen to learn and ready to ask me.'
>
> 'I prefer my new pairing. I find that I know what he's done, whereas I couldn't before. My old oppo. had his own ways of doing things, which he no doubt thinks are right, just as I think mine are, and won't alter them. The change came because there was trouble. Why it was doesn't concern me nor anyone else. It was a bit of both management and union.'
>
> 'I couldn't get on with my oppo. I think it's the difference between a local person and a person from the North. My training was broken into by having to go to the non-automatic department. We were both struggling along – inexperienced. Now I can get help and if my oppo. says he'll do something, I know he *will* do it, and I'll be able to find it when I go to it.'
>
> 'My oppo. was an easy-going chap. I think I was doing all the work. I didn't think it was fair that the two experienced chaps should be together.'
>
> 'That's trade unionism – good or bad, experienced or not – the same wage for all.'

This incident has several important implications. First, the solidarity of the overlookers as a group is clearly displayed by their attempts to solve such a problem within their own ranks. In the case involving the weavers, although a comparable attitude prevailed, the group was not sufficiently organized to take action on such issues. The individual overlookers reported their complaints to their own group leaders, whereas the weavers went straight to the management. Young and Irwin, in particular, exhibited very considerable loyalty to the overlookers' group by agreeing to break up their relationship in order to solve the problems in the other pairs. Most of the group, in fact, accepted the difficulties as the responsibility of the group rather than of the individuals concerned.

Second, only as a last resort did the overlookers approach the management, and then merely to seek authorization of their own solution to the difficulties. The management showed great wisdom in waiting for, and allowing, this natural solution to be reached, for the tension and difficulties were well known to them for some time before the union approached them.

Finally, we can understand why Laycock's earlier offer of an intra-group reshuffle was rejected. At that stage of the innovation the tensions in the pairs were only beginning to appear. Laycock's suggestion that there might be need for a reshuffle was perceived as an attack by him on the solidarity of the overlookers' group. Further, it would have confronted individuals with openly choosing new opposites. Such an action might have been interpreted by the old opposite as stemming from management dissatisfaction, whether it meant that or not. This only emphasizes the difficulty that surrounds the management of such human relationships. Not only must the policy be correct, but also its timing must be such that it meets a need that is acknowledged by the group. Two unanswered questions remain. Would it have been too great a risk to the success of the innovation for the overlookers to have chosen their pairing before the looms began to arrive? Would such pairings have led to the same tension and difficulties?

HUMAN RELATIONS ARE IMPORTANT

The weavers and overlookers who enjoyed good relations with their opposites often commented upon the fact. Its importance was to them a positive concern which led to definite efforts to achieve and maintain the smooth relationship; for example:

'It can be very important, but it all depends on the individual. You've got to work together. People who are all out for themselves will find it a nuisance, because other people won't help them. I've never had any trouble. If anything, my partner's better than I am. We always leave our looms as free from trouble as possible, with the shuttles in the single box side so that they can be oiled easily in the morning. I've never yet come in to find broken weft. If you have a smash you can't help that. . . . We never decided on a policy, but just found that we worked this way.'

'I don't think there could be a better pair than ours for following each other. We work in together perfectly. You have to on shift work, otherwise you spend all your time thinking your opposite is leaving things for you.'

Four of the former non-automatic weavers had already been paired as shift-opposites in their previous department. When they were told of their impending transfer to the automatic section, they asked the manager if they could maintain their relationships on the new looms. He readily agreed to this request. Both they and the manager felt that the understanding and cooperative behaviour they had already established was worth trying to retain. One of these pairs illustrates the sort of adjustment to shift work which can be achieved by good relations between opposites and a management attitude which confers a measure of responsibility for the conduct of the work task on the effective working group – in this case the pair of shift-opposites – rather than the individual weaver. A member of this pair said: 'Well, here I'm fortunate – Greenwood and I have been opposite each other for a long time. We live in the same road and know each other outside the factory – as well as in. We demanded that we be put opposite each other when we went on automatics. He's a keen referee, and I change with him when he wants me to, and he does the same for me. I'm going to night classes once a week and I wouldn't be able to otherwise. We know each other's methods and we've had it out together. We set out to leave things clear for each other. Not that we haven't had rows, but we always get over them and clear the air.'

The management have retained these pairings throughout the numerous reshuffles in the group of weavers. Moreover, examination of the reshuffles referred to in Chapter Four shows that although actual loom setts have altered frequently, the human pairings have been maintained remarkably constant by the management. The latter were as conscious of the importance of the relation as were the operatives. However, not being directly involved in the tension, the managers sometimes failed to support this awareness with action. For example, the principle of providing opposites with the opportunity of actually working together was incorporated in the training programme, but this was not implemented to any extent.

Likewise, the management generally encouraged the opposites to communicate with each other at the change-over of the shifts at 2 p.m. However, no official policy about this overlap was developed nor was any financial compensation made for times spent in this way. One of the shift managers commented: 'Shift-opposites are the secret of this sort of work. You need to leave adequate information – weavers and overlookers, and George and I. We encourage them to leave information at ten and to pass it on at two. Otherwise you don't know where you are, and at 6 a.m. you naturally think that someone is getting at you. I've seen warpers leave quite long notes. It's the first thing you have to detect – trouble between shift oppos.'

Such communication from one shift to the next was essential for maximum efficiency; and, particularly among the overlookers, it was common to observe them together for up to fifteen minutes. The two shift managers themselves often overlapped for a much longer time. The provision of adequate means of communication between the shifts, is a primary problem for the management of such systems to solve at all levels of the social structure.

Nevertheless, extensive communication cannot compensate for

a basic lack of understanding in the relationship, and it was striking that among the weavers, good relations minimized the actual amount of day-to-day communication between the opposites. One weaver said: 'You should see some of the notes some of them leave. If *you* knew where they left them *you* could read them. Long lists of looms with various things wrong and comments. When twenty to ten comes, they start on their lists. We do this only when we can't leave things right.'

The discussion of the relations between a weaver and his overlooker earlier in the chapter can now be extended a little further. We have already seen that the interdependence of these two operatives who occupied different roles in the structure was recognized. Furthermore, some of the weavers commented that this relation of interdependence had to be extended to the other shift as well.

> 'Sometimes it's the other overlooker and not the weaver. The weaver may tell him, but if he won't start the job, the weaver can't do anything about it.'
>
> 'All four have to work together. Our overlookers share the weaving bonus so they want everything to be going, on both shifts. That's the best way. Our group has no trouble.'
>
> 'You could run twenty-four easily on day work. You'd know where you are. But now you get the looms straight and after another shift they are all out of adjustment. This is more affected by the overlookers than by the weavers.'
>
> 'The other overlooker especially affects me. The pair we had before were all right. Even if something happened at seven minutes to ten, they'd do it, but not the others. Again, if I go to him just after two, he says, "Oh, it was all right on the other shift".'

We thus have some expression in the section that the real work group in continuous production goes even beyond that group of different operatives who combine together on any one shift. In fact, it embraces those groups from all the shifts who are linked together by the production from one sett of machines. It is the interrelations of this total work group that are important for the fulfilment of the work task. These groups do not at the moment find expression in the structure and culture of the mill. The effect of such a formal expression of these work relationships on production and on the satisfaction of individuals would be most worthy of experiment and observation. It may throw some light on the perplexing findings of the statistical field studies of Katz (1951a, 1951b) and Argyle (1957, 1958). In four major studies by the research teams under the direction of these authors, a number of variables relating to small groups in industry have been measured and correlated. There was a singular lack of any correlation between productivity and satisfaction variables. These findings are in marked contrast to the clinical observations of such people as Homans (1953) and Rice (1958). The groups in the statistical studies all appear to be role or departmental groups – that is, groups which are defined formally in the social structure of the particular organizations concerned. They do not appear to be work groups as we have defined them, and this distinction between the character of the groups selected for observation, may have an important bearing on the above discrepancy.

SHIFT WORK AND LIFE OUTSIDE THE MILL

The introduction of shift work into the mill had many other human implications when we consider that each individual was involved in another whole network of relationships outside the walls of the mill. The new shift worker was confronted with a range of new situations quite apart from those directly related to his work task. For example, there was no public transport in the Radbourne area for shift workers. Or again, established mealtimes and the pattern of home life required revision and social activities and leisure habits were sometimes affected. All these new situations involve relationships in this other network.

Although some of the operatives in the automatic section had worked on shifts prior to the innovation, an equilibrium pattern for most of the shift employees had not been attained. Shift work was regarded as an inconvenience – a thing to be suffered for a temporary period. The general reaction to it was to modify the established pattern for day work as little as possible, and to hope that this 'unnatural' state would not continue too long. The permanence and possible extension of the present change were not recognized by everybody in the mill. However, the appearance of Gardner and Ashby, as senior staff on shift work, meant to most of those in the automatic section itself that it was a reality which made the above attitude and response untenable. A new response had to be developed.

Of the thirty-three who were in the section during our study, only one operative and his family had had previous experience of shift work over any length of time, and this was in Lancashire, and not in the present area.

The wives of twelve of the fourteen married men in the section were interviewed at home concerning the effects of shift work on their home and family life. All of them found that their usual pattern of living had been markedly altered. Just half of these wives emphatically wished their husbands were back on day work. The other half had found a number of advantages in the new pattern, which largely offset its disadvantages as far as they were concerned. Their attitude had only been attained gradually over the months of experience of shift work. Some of them had earlier expressed the strongly unfavourable opinions about shift work that characterized the first group. The attainment of a favourable attitude to shifts was encouraged by the inevitability of these conditions of work for their husbands. Radbourne was only a small town and alternative employment was non-existent, especially for the trained personnel. The mobility of families in the area was not great, and only one family even contemplated moving when shift work was mooted.

This gradual attainment of a degree of acceptance of shift working was also typical of most of the operatives in the section. The married women in the mill universally found the obvious domestic advantages of shift work to their liking and they were all strongly in favour of the new conditions. For example, they were now able to fit in their shopping and housework during the part of the day when they were not at work. They also found that they were less tired on shifts than on days. However, this difference in tiredness may have been due to the change in work task, from weaver to spare-weaver, that most of these women experienced when they began shift work.

The married men in the mill, on the whole, found fewer advantages than disadvantages. Most of them appreciated having some leisure during the day, especially in the winter when they formerly saw little daylight. This leisure time was spent in gardening, hobbies such as bird fancying, and doing jobs about the house. However, this extra leisure time was not so compelling an advantage for the men, who already had two days at the weekend, as it was for the women mentioned above. On the other hand, the men disliked the changeable character of the work, the irregular meals, the disruption of regular evening activities, and complained of greater tiredness on shifts. None of these complaints was independent of relational problems. All were complemented by opinions from other individuals who were in a position of relationship with the men. For example, irregular working, when enlarged upon, led to the problems with shift-opposites. Second, the wives of these men disliked having to prepare meals at extra times of the day, for children at school and husbands on shifts. Again, regular evening meetings which formerly provided important social contacts with friends or relatives had now to be given up. Finally, both husbands and wives complained of the problem of getting adequate sleep under a shift system. On the early-rising week, there was a tendency to stay up late, because it was the only week with any evening leisure time. The other week did not allow adequate compensation, because there was still the necessity for rising early to get children off to school. In practice, both husbands and wives had appreciably less sleep on shifts than they had had on days.

The non-working wives found advantages in having their husbands at home during the day, and commented:

> 'Shopping is easier, because either my husband can do it or else he can mind the children.'
> 'My husband can help me more in the house.'
> 'It's nice to have my husband home in the afternoons.'

Nevertheless, this new situation with its air of permanence meant that husbands and wives were confronted with new adjustments. Particularly for the husbands, the adjustment in their role entailing greater participation in the care of the children and in the home was not always easy. The main disadvantage, apart from the extra meals, for the non-working wives was that they were left alone at night every second week. It is of interest to note that television played a not unimportant role in alleviating this aloneness: 'It's just like having someone with you.'

Two families in particular remained very opposed to shift work even after a year's experience of it. These cases differed from those described above, whose opposition stemmed from a reluctance to change existing modes of behaviour. The second group of couples readily altered their established modes to fit in with the new requirements. One of the husbands cycled home during the short meal break on shifts, even though he was there for only ten minutes. It was also notable that these alterations were not the object of the unfavourable comment in these two cases, as they were for most others. These couples disliked the absolute limitation that shift work imposed on the time they could spend together with their children. Under the shift system, if the children were at school, they had virtually no contact with their father every second week. These couples made the best of the possibilities of shift work, but the value they placed on a shared family life was such that shift work was still disliked.

The single people employed in the mill were generally opposed to shift conditions. They complained both of boredom with the day-time leisure and that their nights were not free.

These observations on the effects of shift work outside the immediate work situation, emphasize the importance of considering the wider social field of individuals in any studies of limited social systems such as the factories in our study. This was very evident in the cases of the operatives for whom the changes to shift work and the looms occurred simultaneously, and they said:

> 'I don't mind the idea of learning the new looms. They are machines and can be made to go like any other machine. Since they are precision machines, they should go better. What does worry me is the prospect of shift work. I can see shifts are coming, but I don't like the idea. It's the changes in my general life that bother me.'
>
> 'I wasn't too sure about shifts so I hesitated several times when I was asked to go in the section. I would like to have tried it for a while, but I knew that once I was on shifts, I would not be able to change again.'
>
> 'Shifts are the whole trouble. I have been asked whether I'm satisfied with the money, and of course, everyone will take a little more – but that wasn't the point. Shifts just don't suit me. They may be some people's cup of tea, and getting meals may be a little thing, but it just doesn't work out for me. . . . I like the work and these looms are easier than the non-autos. Shifts just interfere with our marriage.'

Towards the end of our study, the management made a tentative suggestion of a three-shift system, but this was met with organized resistance by both unions. Most of the operatives perceived such a system to involve much more serious social effects than those of their present double-shift system.

SUMMING UP

In this chapter we have traced the specifically social changes arising out of the new working conditions. Basically these changes were derived from two very different sources. In the first place, there were the different relationships which resulted from weavers working on much larger numbers of looms. This altered the amount of time they could spend with different workers in the mill. In the second place, double-shift working brought into prominence relationships *across* shifts – a change that was not a direct result of the technology but stemmed from the inevitable sharing of responsibility for the machine's production.

We have also described the concept of a 'work group' and have noted that, however rationally desirable the realization of such groups may be, there are many factors in the mill which operate against this possibility; in particular the prominence of the role group, which is reinforced both by official recognition and also to some extent by union membership.

Finally, we have discussed some of the effects which the changes in the mill have had on the relationships of the workers in their domestic and spare-time activities outside the mill.

REFERENCES

ARGYLE, M. *et al.* (1957). The measurement of supervisory methods. *Hum. Relat.*, vol. X, pp. 295–313.

ARGYLE, M. *et al.* (1958). Supervisory methods related to productivity, absenteeism and labour turnover. *Hum. Relat.*, vol. XI, pp. 23–40.

HOMANS, G. C. (1953). Status among clerical workers. *Hum. Organization*, vol. 12, pp. 5–10.

KATZ, D. *et al.* (1951a). *Productivity, supervision and morale in an office situation.* Survey Research Centre, Ann Arbor: University of Michigan.

KATZ, D. *et al.* (1951b). *Productivity, supervision and morale among railroad workers.* Survey Research Centre, Ann Arbor: University of Michigan.

RICE, A. K. (1958). *Productivity and social organization: the Ahmedabad experiment.* London: Tavistock Publications.

Works not specifically referred to in the text

HERBST, P. G. (1962). *Autonomous group functioning*. London: Tavistock Publications.

MANN, F. C. & HOFFMAN, L. R. (1960). *Automation and the worker*. New York: Holt.

TRIST, E. L., HIGGIN, G. W., MURRAY, H., & POLLOCK, A. B. (1963). *Organizational choice*. London: Tavistock Publications.

WALKER, C. R. (1957). *Toward the automatic factory*. New Haven: Yale University Press.

NINE

Communication

'Social organization without communication is impossible', says Miller in *Language and Communication* (1951), and although this is undoubtedly a truism, it is nevertheless of fundamental importance in understanding the operation of a social group.

In our particular context of industrial social groups, 'communication', like 'human relations', has become a jargon term in recent years, and many attempts have been made to improve or install aids to communication.

It was clear that in our approach to industrial change, information – which is the subject-matter of communicative acts – is of vital importance at varying levels of social operation and with varying degrees of complexity.

If it is borne in mind that information is the central part of communication, then it is obvious that we extend an inquiry about communication much more widely than might be supposed by the meaning currently given to the term. For information is vital to all members of an industrial social group at all stages of their work, and information that is derived from many sources other than such official ones as the notice board or the works magazine.

Verbal contact in a person-to-person meeting is an obvious means of communication, but in the broadest sense, any interaction between individuals is a communicative act in that each person will either receive information of some sort or give information – or both.

In practice, it becomes essential for some communicative acts to be formalized into some sort of system such as a progress bulletin or an order slip, and we shall call this *non-personal communication.* This streamlines the process of communication and, by establishing a constant frame of reference for each communicative act, enables the object of the communication to proceed with the minimum of distortion.

However, even under these conditions, it is possible for the purpose and meaning of the communication to be distorted even though the content of the communication cannot be affected. For example, if one department receives a large number of official demands from another department, the recipients may feel that the originators of the requests are doing their best to try to show up the deficiencies of the supply department. Therefore, it is true to say that systematic and formal communication is still only *relatively* free from distorting influences.

In a factory there are some additional complicating factors that make the situation in which communicative acts occur far more complex than the laboratory situations which have been used from time to time in the study of communication. First, there is almost certainly a minimal level of communication which must be reached in order for the organization to function at all. Second, there is the superior–subordinate social structure of industry which specifies and controls the socially allowable sorts of communication, and these in turn operate according to the specific culture of the organization. It may, for example, be quite permissible for an unskilled man working in the factory to speak directly to the head of the organization, but this mode of communication might be quite impossible in the case of a junior supervisor.

Thus, because an industrial organization is functional (i.e. exists for the production of goods) and because it is in some degree hierarchical (i.e. arranged according to superior-subordinate positions), the pattern of communication is of necessity keyed to these two factors, and must be considered in these terms.

COMMUNICATION IN THE MILLS

For the purposes of this particular study, communication was chosen as being a topic of especial interest in relation to the work people of the mill. For many of them, communication and information – although important – were still only contributory to their main work, whether this was weaving or some other task. For the supervisors or managers, however, communication played a much more important part in their job function, and therefore we have considered it in some detail in Chapter Five. As we saw there, some members of this group spent almost their whole time gathering and disseminating information. We shall, then, of necessity refer to communication in the management group, but our main interest will be with the operatives in the mills.

A further important point concerning our subject material in this chapter is that much of the study of communication which we made was not longitudinal. That is, we did not attempt to track the development of communication as the change in the mill progressed, but rather sampled the existing communication in both mills over a given period of time, using the operatives themselves as indicators of changes.

In studying communication in both mills, we were really working on an assumption. This assumption was that, because the mill at Debenham was a stable and efficiently running unit, we could assume that the communication system there was, at the very least, satisfactory. Against this standard, therefore, we hoped to be able to compare communication at Radbourne.

We were fully alive to the dangers of this procedure, and in the analysis that follows it is hoped that its limitations as a comparative study are made explicit.

Types of Information

Although it is an analytical concept and not related to actual communicating behaviour, it is useful to say something about the types of information which were found to be relevant to the working life of the operatives.

First, there was information relating to the immediate working task. (Type 1 information). This included the supply of materials, criticisms and complaints about the work given to and received from the supervisor, and aspects of work-study information relating to the immediate work.

Second, there was information relating to the immediate function of the department or section of which an operative was a member (Type 2 information). This was concerned with interdepartmental communication and with communication between shifts where this was appropriate.

Third, there was information relating to the functioning of the mill as a whole rather than to the operatives' immediate department (Type 3 information). This concerned the general production programme, the work of departments other than those with which the work people were connected, the general commercial activities of the company, and the place of the industry in the national economy.

Information of the first type is almost inherently necessary in order that work can be done at all. Information of the third type is the sort that provides the target for most of the industrial and political concern over 'communication'. But information of the second type is more nebulous and less well attended to as an industrial problem.

Because it was our working assumption that the mill must be treated as a total social system whose separate departments are all interdependent, we gathered information concerning communication from a sample of all the work people in each mill. That is to say, although the innovation we observed took place in one section of the mill alone, the effect of this would, we felt, be experienced throughout the total field of the mill's activity.

In this respect, the activities of the mills can be divided into two broad categories – the weaving activity and the non-weaving activity. In many ways it is implicitly held in the mills – and explicitly expressed in numerous status symbols – that the weaving function is the 'real' function of the mill. The ante- and post-weaving processes, which include the preparation and processing of the yarns, the inspection of the cloth, and so on, are all perceived as more or less ancillary functions that serve the weaving section. The evidence from wage-levels supports this view, since weavers and overlookers (loom mechanics) are the two most highly paid groups of work people. In addition, the senior positions in the management group are occupied by people who were previously weaving personnel.

We have maintained this division as being socially relevant to the description of communication despite the technological fallacy inherent in it.

Methods of Communication

Communication in any organization or social system tends to be of two classes which are roughly analogous to the concepts of 'formal' and 'informal' social structures. The formal structure is considered as the codified system of job relationships which is the theoretical basis of the working of the organization. It may or may not be set out in symbolic form, but it is generally known to exist. The informal structure, on the other hand, is the actual existing system of face-to-face relationships which are certainly structured but, as opposed to the codified form, are not regarded by the participants as existent for the production purpose of the organization. This is never set down, but is implicitly recognized by all the participating individuals. This structure may or may not be conterminous with the formal structure. In practice the face-to-face groups tend to lie within the job groups and horizontal levels which appear in the formal structure. We have discussed in Chapter Six the effect the new looms had on the nature of these relationships and the groups in which people in the automatic section found themselves.

In a parallel manner, we can describe one class of communicative acts as formal and explicitly defined. An assumption here is that where the need for communication is consistent, and also either continuous or regular, then such a systematic method of communication will develop. These so-called formal channels of communication will develop in such a way as to conform to the pattern defined by the formal social structure.

On the other hand, there is communication which is quite informal in character, in that it is in no sense institutionalized, but nevertheless plays a vital part in the communication linkage.

Communication Patterns at Debenham

At Debenham most of the communication to the operatives was effected by person-to-person contacts. The information was derived from the immediate supervisor, and upwards communication was also generally referred through him. 'We make it our business to find out [about work] from the foreman.' 'When we finish the job, we just go and ask Jack for the next.'

There were two broad exceptions to this completely personal means of communication. One was the Works Council minutes, which naturally dealt mostly with information of the third type outlined above, and they were available by being passed from hand to hand. Thus the channel was both personal and non-personal in character. Although it was almost inevitable that someone or other would not see the minutes for a particular month, nevertheless most work people read them fairly regularly.

The other main exception to personal communication was in the weaving department. Here, most of the operatives received information about work – at least in its main aspects – from a book which was located on the supervisor's desk. This book indicated what sort of cloth was to be woven in which loom. It also gave some details of the type of yarn which it was proposed to use, and whether this differed for warp and weft yarns, or whether they were identical. In one way, it might appear that the use of written communication in the weaving shed was almost inevitable since the very high noise-level made conversation arduous. However, the operatives were very adept at overcoming this barrier to conversation – a fact which surprised us as observers.

As far as formal personal communication was concerned, there was also the Works Council, which met regularly at monthly intervals and included some operatives as representatives. The reader will recall from Chapter Three that there was a yarn processing unit at Debenham as well as a weaving unit, and that each unit had a separate manager. The Works Council dealt with both these units and was under the chairmanship of the yarn processing manager, who was the senior manager of the two. At this meeting there was some personal communication of information – of the third type. The Works Council at Debenham – because of its shared nature – inevitably dealt less with interdepartmental information than did the Council at Radbourne.

It remains true, however, to say that communication in the mill at Debenham was characteristically of the personal, sporadic kind already discussed, and was essentially a two-way process. That is to say, communication went as easily up the hierarchy as it came down.

Since there were virtually only three levels in the work hierarchy, distortion of information in either direction was really minimal, and, in addition, information in all three fields was continually being distributed at the daily production meeting (see p. 34) and so inevitably filtered down fairly rapidly to the work people themselves. Having reached the operative level, it then spread throughout that level by means of interactions arising from the task, the layout of the department, and the informal social structure.

Such operative-to-operative communication was most important for the shift workers. Here, as the people concerned repeated to us time and again, it was vital to communicate with the person doing the same job on the opposite shift. There was no officially allowed time for this exchange of information; the exchange could only occur by the work people themselves overlapping for a short time on the shift change. Despite the lack of a formal channel between shift-opposites, such communication was explicitly and implicitly assumed by all.

Communication between operatives was also fairly free during their working time together. Because of the small size of the unit, and lack of physical barriers, most workers were able to interact fairly freely with their fellow-workers if they chose to do so, and there were, of course, the task-centred interactions already referred to.

The main physical barrier in the mill (see *Figure 1*, p. 27) was that between the preparatory sections and the weaving shed. Interaction during work between weavers and preparatory workers was virtually non-existent. Nevertheless, there was a third group of ancillary workers who effected some sort of link between the two main divisions. These people were responsible for conveying materials from one place to the other and doing ancillary work on the looms themselves. Their roles as key persons in the informal channels of communication were most important.

In sum, operative communication in this efficient automatic unit was characterized by a free flow of information, mainly – but not completely – disseminated by personal, occasional interaction between operatives and management and between operative and operative.

Communication Patterns at Radbourne

Reference to *Figure 4*, p. 41 will indicate that the departments of the mill at Radbourne were much more discrete than those at Debenham. In addition to this, the management group itself was much larger, and consequently each process tended to have its own supervisor. These two facts are important in considering the different aspects of communication.

As in the case of Debenham, it is true to say that much of the important communication in the mill involved occasional personal interaction between work people and supervisors. There were some exceptions to this, just as there were exceptions in the Debenham case. First, there was the communication of the Works Council information. Officially, this was made available in the form of summaries of the meetings, which were placed on the general notice-boards scattered throughout the factory. Second, in the entering department, some use was made of work schedules, so that although the person requiring more work had to get *additional* information from the supervisor, a limited amount of information was available from written records, which were public. It is interesting to note, that in this department the foreman took responsibility for organizing the work of a group of 'knotters' and their assistants. Their work was concerned with the warps, but was done actually on the loom in the weaving department, and not in the entering department. Thus for much of their working time they had no direct supervisory contact.

The last partial exception was in the automatic loom weaving section. Here, towards the end of the study, a copy of the forecast of work for the next week was posted up, but there was some evidence that this was not used as a reliable communication source by all the operatives at that time – partly because its existence in the section was not known by everybody.

As at Debenham, the importance of good inter-shift communication was stressed by most of the work people, and not only for communicating information about the immediate working situation. Other aspects of work are as important as technical information concerning the work itself. An illuminating comment on this aspect was the following: 'As well as work, shift information is important from the point of view of knowing how things stand with the foreman.'

Figure 9, p. 64 indicates that there was increased differentiation of function at Radbourne, and with this factor went also a more rigid barrier between departments. As in the case of Debenham, there were a few interdepartmental operatives, but for the most part operative-to-operative communication between departments was virtually non-existent in working hours.

Work-Study Communication – a Special Case

Information concerning work-study was central to the actual work of the operatives, but there were certain peculiarities in relation to it, which deserve closer study here, more especially in relation to weaving operatives.

We have already given the basic facts of the application of work-study in Chapter Two, and the reader will readily appreciate that the schemes were simple neither to apply nor to understand. Since the whole area of the operative's task and his reward for performing it were largely determined by work-study, the communication of work-study information was of great importance.

The pattern of communication for work-study was similar at both Debenham and Radbourne. The method was primarily a formal personal channel between the work-study engineer and the operative and more often than not the immediate supervisor was bypassed. A typical comment from an operative was: 'If I have a complaint I go straight to Jack Moore [work-study engineer] myself.'

This pattern was in use in both mills, but at Debenham it was much simpler to operate because the work-study office was adjacent to both weaving and preparation sections of the mill. At Radbourne, the office was a considerable distance from all departments, especially the weaving department, and physically separate from the production area.

Our particular interest in work-study communication was that it was the sole example of specialist communication. We have noted that on the whole Type 1 information was routed through the supervisor either by informal, personal means or by formal impersonal means. The reasons for the present exception seemed to derive primarily from resistance to work-study by both operatives *and* supervisors, particularly in the weaving sections where the schemes were probably most complex. In other words, the supervisors refused to be the communication link in this instance and thus refused to be implicated in the decisions which were taken as a result of work-study methods. The weaving supervisors in particular felt strongly and explicitly about the scheme – as one said: 'The whole scheme is shaky. If I was a weaver, I'd be at [the work-study engineer] all the time over various points.'

The difficulties which inevitably arose over the scheme were therefore forced to be taken directly to the work-study department.

We were able to observe, too, that many weavers on entering the new automatic section at Radbourne were given no explanation of the scheme, and the responsibility for such explanation was not clearly defined.

Although work-study communication was very similar at both mills, the coherent supervisory system at Debenham – coupled with the physical proximity to the production area of the work-study office – enabled many of the possible difficulties to be quickly ironed out. These conditions did not obtain at Radbourne, and more frequent disagreements were probably only avoided by the liberal personality of the work-study engineer and his capable handling of the disputes that arose.

There is no doubt that, since the work-study techniques were not completely acceptable to either supervisors or others in the weaving sections, the communication channel was not integrated with the rest of the system.

ATTITUDES TO COMMUNICATION

We have now dealt – in brief – with three main concepts:

(*a*) the *types* of communication
(*b*) the *modes* of communication
(*c*) the actual communication *usage* in each mill.

How did the workpeople feel about communication? Was it adequate or not? And are there any peculiarities of the automatic system which affect communication? Do such peculiarities differ for the people engaged in weaving and non-weaving activities?

We shall first consider these questions in relation to each mill separately. The attitudes of the non-weaving operatives to the existing pattern of communication at the two mills are given in *Tables 11* and *12*.

TABLE 11 DEBENHAM MILL:
NON-WEAVING OPERATIVES' ATTITUDE TO COMMUNICATION

Communication Type	*Communication Method*	*Workers' Appraisal*
1	Informal, personal (mainly through supervisor)	Quite satisfactory
2	Informal, personal (through supervisor and other operatives)	Satisfactory, although not important for most people
3	Formal, personal/non-personal	Satisfactory for some; insufficient information for others

TABLE 12 RADBOURNE MILL:
NON-WEAVING OPERATIVES' ATTITUDE TO COMMUNICATION

Communication Type	*Communication Method*	*Workers' Appraisal*
1	Mostly occasional, personal through supervisor; in entering department, some systematic non-personal	Satisfactory for weft preparation; mostly satisfactory, but sometimes not, in warp preparation
2	Occasional, personal (through supervisor and ancillary workers)	Of no importance
3	Formal, personal	In weft preparation, no interest; in warp preparation, some desire for more information

In comparing these two summaries, one of the first things which is notable is that there were differences between the different departments at Radbourne which were not apparent at Debenham.

This is probably mainly due to the fact that *all* preparation at Debenham is under one main supervisor, and, what is more, although not very cohesive, the workers share some measure of belonging to a common group, i.e. 'preparation department'. This is facilitated by the lack of physical barriers at Debenham. In fact the main physical barrier in the mill – a wall between the weaving and the preparatory sections – helps to define this area as the 'preparation department'.

On the other hand, at Radbourne each department was traditionally distinct and separate having its own particular location. It is still within the memory of some workers when the warping, in particular, was a secret process, and only those actually in the department were allowed to have knowledge of it.

Two illustrations of this separateness will clarify the situation. One of the interviews conducted was with a mobile worker whose job was to transport semi-finished raw material from one preparatory department to another. At the time of the interview, Radbourne Mill was working at full pressure, and as a result stocks of semi-processed material were rather large in every department. The complaint of the particular worker was that, although his own department was turning out the material rapidly, when the stocks of the receiving department had reached a certain level, the supervisor of this department refused to take any more material from him – although the worker himself could not 'stockpile' in his own department because of lack of space.

The second illustration concerns the shift conditions. One particular worker had jobs in two distinct preparatory departments during shift hours, although he worked in one department only during day-hours. Because of pressure of work from both departments in shift hours he found it necessary to decide which department had priority. However, he did not refer this problem to the shift manager – who could have been expected to deal with it since it only arose in shift hours – but rather referred it to his day supervisor, who in turn asked the assistant manager to decide the priority.

In general, it was found that the increased pressure of automatic working had not altered operative behaviour to any degree in the preparatory departments. The personal type of communication was still adhered to in both mills, and there was nothing in the preparatory processes for automatic working which could not be simply incorporated in the skills of the operatives. We have already argued in Chapter Five that the pressure of communication was most felt by the supervisors and that the preparatory supervisors 'took the strain', as it were, of the changes which the new technology brought about. This is almost entirely true, except for the sizing process in the preparation of warps. In this department the pressure for warps for the automatic looms was felt by the operatives at Radbourne from time to time. As one of these operatives put it: 'They always want you to produce more and more. It's not really a personal pressure but just a demand for more warps.'

The two main changes for preparatory operatives that arose as a result of the introduction of automatic looms were (i) change in the quality of the work required, (ii) change in the quantity of the work required. From the evidence concerning communication, it seems that the preparatory workers in both mills were able to produce the quality of work which automatic looms required without any change in the communication pattern, and without the introduction of any real means of sanction for quality. Quantity of work is, as has already been suggested in an earlier chapter, largely the responsibility of the supervisor, and here again the preparatory workers generally had adequate information from him within the existing communication system.

In more theoretical terms, it seems that the work-culture within the preparatory departments was sufficient as it stood to absorb the demands made by a more automatic technology, although this is not necessarily true of the supervisors' group with their new interdepartmental activities. We have also seen, however, that the smaller size of the preparatory unit at Debenham aided communication between its sections but that the culture pattern concerning communication was not essentially different from that at Radbourne. Both depended formally on personal transfer of information between a supervisor and a single operative. The only group dissemination of information was by means of the informal channels of more face-to-face groups which existed.

The next consideration is that of the weaving workers at both mills, and a tabular summary of their attitudes is as follows:

TABLE 13 DEBENHAM MILL: WEAVING OPERATIVES' ATTITUDE TO COMMUNICATION

Communication Type	*Communication Method*	*Workers' Appraisal*
1	Formal, non-personal *and* informal, personal (through supervisor)	Quite satisfactory
2	Informal, personal	Sometimes inadequate although generally satisfactory
3	Formal, non-personal; informal, personal	Not enough information

As we have already described, at Debenham the weaving workers made considerable informal use of records kept by the supervisor. From these they were able to obtain information concerning future work on particular looms without further reference to the supervisor except in cases of sudden change. This system worked well and was used by most people.

At Radbourne – where there was not such a plentiful supply of

TABLE 14 RADBOURNE MILL: WEAVING OPERATIVES' ATTITUDE TO COMMUNICATION

Communication Type	*Communication Method*	*Workers' Appraisal*
1	Informal, personal *and* some non-personal	Not adequate
2	Sporadic, personal	Generally adequate
3	Systematic, non-personal	Some desire for more information

information – the work people felt the need of more information of an immediate working type. The extreme point of view was expressed by one worker: 'The management don't really seem to tell us a lot about the work – you have to find out for yourself if you can.'

From both mills there were requests for more information about the finished cloth (i.e. the cloth ready for the retail market), preferably given in a non-personal, formal manner, as by examples of cloth displayed or formally distributed in some manner.

In Chapter Six we have discussed the overlooker–weaver relationship in some detail, and there was some evidence that the communication relationship between these two workers differed between the mills. For example, at Debenham one overlooker said: 'The weavers shouldn't go direct to the supervisor although sometimes they do. The overlooker is sort of the go-between for the weaver and the supervisor.'

This was an extreme view rather than a norm of conduct, but other operatives – both weavers and overlookers – expressed a preference for this system.

Because the weaver is so much more tied to a geographical location than the overlooker, it was almost inevitable that the overlooker would develop communication contacts more readily than the weaver. Freedom of movement in the operatives' role – as in those of the overlookers and the interdepartmental personnel – is an important requirement when looking for key personnel in the informal channel of communication.

In general, and comparing the two weaving sheds, it is true to say that at Debenham there was satisfaction with much of the present communication of information, but also some demand for increased information about company activities, finished cloth, and so on. One person, commenting on the general situation, said: 'We get quite enough information. In fact, you know, this is generally a much better shed [for that] than most.'

At Radbourne there was both a desire for more information about the wider aspects of the job and some dissatisfaction with the communication of information of our first type, that is the actual job information itself.

It is perhaps important to stress that each warp always carried a card with some technical information on it. The dissatisfaction concerned *additional* information that would be helpful to the workers in the performance of their task.

The pattern at both mills was similar, but – especially in communicating Type I information – the Radbourne system was more erratic. The operatives there expressed a much greater feeling of having to *seek* information rather than being provided with it. At Debenham both supervisors and operatives felt that the supervisor had functions in relation to operatives wider than the mere dissemination of work information. His job was rather to ensure the smooth running of the department and to make sure that the information and working material were provided and readily available.

The operatives at Radbourne were not simply grumbling diffusely because of the difficulties of working in a new section, but because they needed a communicating system at least similar to that established at Debenham in order to meet the minimal demands of continuous weaving. This was exemplified by one of the Radbourne shift managers, who said: 'There is more pressure from work people on autos . . . and so you've got to act proportionately quicker.'

One special part of the communication of Type 1 information deserves more attention. This was the system by which faulty cloth was notified to the weaver and overlooker.

Woven 'packets' of cloth were examined by cloth examiners and graded according to the number of faults found. If the grade fell below a certain standard, a ticket was hung on the loom which had produced the low-grade cloth. This part of the system was followed at both mills. In addition to this, the weaver and overlooker were called separately to the cloth inspector and shown the damaged cloth. Following this, the loom was supposed to be stopped at the next packet until the cloth had been inspected.

There were interesting differences between the two mills in the way in which this formal system worked. At Debenham, the 'damage ticket' was hung on the loom, but the other parts of the inspection system were often not followed through. Sometimes the weaver and overlooker saw the cloth, sometimes they did not, and the loom was rarely stopped at the next packet. This did *not* mean that the problems of producing good-quality cloth were not attended to – they were, but they tended to be a major problem of the senior supervisor, Morris, who said: 'My job is primarily that of watching quality.'

This conception of his role was confirmed by the inspection supervisor and by the weavers and overlookers themselves. It seemed to us that this system of quality control was entirely in harmony with the demands of the technology in that the continuity of the looms was ensured, and the responsibility for bad quality was felt to be a shared responsibility rather than an individual fault. The cloth inspection was, as it were, a service to the weaving function – a fact revealed by one weaver who said: 'I go and see faulty cloth if I have time.'

This system of communication was entirely informal and seemed to have developed out of the automatic technology.

At Radbourne the damaged cloth system was followed right

through, including inspecting the damaged packet and often stopping the loom.

But this was clearly perceived by some operatives as being contrary to automatic technology, as we have already noted in another context (see p. 112). The operatives felt strongly that the system was outmoded because (*a*) it took the weaver away from the looms when he should be supervising, (*b*) large setts of looms presented quite different quality control problems from two- or four-loom setts, (*c*) the overlooker–weaver–loom partnership was much more integrated than in the non-automatic case.

For all these reasons, many operatives felt that the loom-stopping and cloth inspection were an anachronism for automatic looms. However, there was no evidence that a modified system was being developed at Radbourne as at Debenham, and indeed this might well be particularly difficult in view of the mixed technology in the mill with both non-automatic and automatic looms side by side.

COMMUNICATION AND CHANGING TECHNOLOGY

As a result of our initial study in the mills, we approached our analysis of the communication practice with two assumptions which we also applied to other aspects of the changing conditions in the mill.

1. The increasingly automatic technology *would* change communication practice.

2. This would not necessarily be confined to the department in which the innovation occurred but would also spread to the other departments of the mill.

Reverting to our theoretical analysis of this situation, we can say that because the frame of reference provided for the worker by the techniques he employs has changed, so will communicating behaviour – which is an outcome of a particular frame of reference. Two immediate characteristics of automatic weaving technology were discussed in Chapter Three. These were:

(*a*) A greater quantity of cloth is woven more continuously.

(*b*) The quality of preparation processes must be improved and related directly to the requirements of automatic weaving.

As far as communication is concerned, therefore, it seems that the first characteristic is the one which most affected operative communication. With the second characteristic, it may mean setting slightly higher levels of skill, and more attention to hitherto unattended parts of the preparatory processes, but there is a reserve in existing practice to accomplish these ends.

It is in the continuity of automatic working that communication problems seem to arise, and in a marked degree for weaving workers. From our analysis earlier in this chapter it seems that, in this particular type of automatic technology, workers other than those on the weaving tasks were *not* greatly affected as regards communication. We have already argued that for the supervisors of the non-weaving sections this is a different matter altogether, but the workers in these sections continue to work in much the same culture – and thus with much the same communication patterns – as before.

Returning now to the working assumptions stated at the beginning of this section, we can say that, as far as communication was concerned, the automatic technology we observed did seem to bring about an increased demand for some types of information from those workers intimately concerned with the looms. However, this did not spread to those who worked on other aspects of production – though one must remember that the supervisors of these sections faced different problems. It is possible that, had the automatic innovations been more drastic or more unorthodox, the communication needs of the other departments not immediately affected would have altered.

COMMUNICATION UPWARDS

The burden of this chapter so far has been communication about the various areas of work travelling from management and supervisors to the operatives, or 'downwards' communication as it is often described. But what about the complementary process of communicating upwards? Clearly, there are many occasions on which the worker wishes to communicate a complaint or a comment about the working situation to those superior to him.

At both mills, the commonly accepted channel for this type of communication was through the immediate supervisor. He would normally handle all such communication, and it was only in exceptional cases that a worker would use either of the two alternative formal channels open to him. These other channels are: (i) the union representative, and (ii) the Works Council representative.

Generally speaking, throughout both mills, the personal sporadic communication through the immediate supervisor was quite adequate from the point of view of the workers, who felt that their supervisors dealt effectively with their complaints and comments. There was one exception to this, and this was the new automatic section at Radbourne Mill. Here there were indications that this mode of communication was not entirely adequate, and several comments like the following emerged during our interviews:

> 'I tell the supervisor and then it's his business.'
>
> 'I tell the supervisor [about complaint]. It doesn't mean you get anything done about it, but he knows.'

Can we find any reasons for this gap in what seems to be otherwise an adequate mode of communication? There are two possible factors which could contribute to an explanation of these observations.

First, the function of the supervisor was different in the automatic case. In Chapter Five the role of Gardner and Ashby was explored in more detail, but for the present point, the important thing is that they were both direct supervisors of the new automatic section and also general shift managers. Every other section of looms in the mill had its own direct supervisor whose role was largely connected with his section. It thus follows almost inevitably that, as shift managers, these people would be proportionately less available for carrying out what was regarded as a direct supervising task by the operatives in this section. Indeed, this is supported by the fact that, for a time, a non-automatic weaving supervisor was given a watching brief over the now automatic section as far as quality of cloth was concerned. Concerning this point, one of the shift managers said, 'A supervisor on the new automatics would be a help because we can't be everywhere at once.'

The importance of the direct supervisor for this task was also shown in another incident at Radbourne. Referring again to Chapter Five, we have noted there the appointment of Black as joint supervisor for the sizing and warping departments. Under him in the sizing department were two charge-hands who had been in the department for some time. In the warping department, he left no subordinate. Inevitably, he was taken up with the problems of sizing, but this had adverse effects on the warping workers as far as information was concerned. This situation was eventually remedied by the appointment of a probationary foreman, but only after some months of Black's new appointment. The accepted pattern of communication in this mill thus seemed to call for a supervisor in close and regular touch with the operatives.

These changes in management structure which were so suited to the other demands of the automatic technology failed in this aspect of communications because the culture of the mill did not alter at the time of the changes in social structure.

The second explanatory factor lay in the perception by the weaving operatives of the importance of the link between themselves and the supervisor. They undoubtedly perceived this link as of enhanced importance in the automatic case. And this was emphasized by the shift managers themselves, one of whom said: 'There is more pressure from work people on autos . . . and so you've got to act proportionately quicker.'

Thus the automatic weaving workers demanded the presence of a supervisor in the same way as other sections in the mill did, but also more active communication than did other sections. The inadequacy of the changes to cope with this demand led to an upward pressure from the section to the immediate supervisor. This effect is similar (though in a reverse direction) to the diffuse management pressure experienced by the sizing operatives. Specific communication items, if they do not result in appropriate action in person-to-person links, mount up to a general feeling of pressure or tension.

THE WORKS COUNCIL

Finally, we must examine in greater detail the Works Council as a particularly structured type of formal communication.

Shortly after the end of the war, Works Councils were formed at each unit in the company as part of deliberate company policy, and their stated aims were as follows:

1. To provide direct management–worker contact.
2. To consider jointly matters affecting the whole mill.
3. To cooperate together to promote the well-being and efficiency of all.

The Councils are composed of elected members who hold their seats by proposal and secret ballot held in the sections which they represent, and other members proposed by the manager from among the supervisory group.

The Councils were excluded from discussing anything bearing on wages or conditions of employment, which were the prerogative of the unions in negotiation with the management.

Thus, in fact, the Council did two rather different jobs. It acted as a forum for the expression of ideas and information from both management and workers, and it also had some executive power in the arrangement of 'social' and entertainment activities of one sort and another. From an immediate point of view, it is the first, rather than the second function which is of importance here.

From the operatives' point of view, the communication provided by the Works Council was basically that relating to Types 2 and 3 of the total communications as we defined them at the beginning of the chapter. That is, it was peripheral to the immediate working task, although it was often the only source of communication concerning planned or impending changes.

As far as the operative was concerned, information from the Council reached him either through his representative or some other person, or through written information about the proceedings of the Council. At Debenham the minutes of the Council were passed round so that, on the whole, most people saw them. At Radbourne, however, the practice was to pin up summaries of the meeting at various points through the mill.

At Debenham, of those people we interviewed most saw the minutes, whereas at Radbourne half those interviewed reported that they never read the summaries, and only one-fifth said that they did so often. At neither mill was there any indication that the Council was perceived as central to the working task, although very few were antagonistic to it as such. The large majority at both mills thought the Council was worth while and should be continued. This last statement might seem to be a contradiction in terms of the previous statements, at least in the Radbourne case where so few people knew about the work of the Council. However, although the activities of the Council – other than those of arranging social functions – were not widely understood, the feeling of 'it's-nice-to-have-it' was strong. This diffuse support was in keeping with the progressive tradition in the company, and management and operatives viewed the Council as enlightened policy.

If one of the functions of the Council is to act as a 'pool of information', as we have suggested, then our evidence is that probably neither Council was adequately fulfilling its task. *Tables 11–14* (pp. 190–194) make it clear that in both weaving and non-weaving sections communications of Type 3 were felt to be inadequate. And again it is in just this area of communication that the Council probably did most of its work.

The reasons for the breakdown were not clear. However, of several factors contributing to it, probably one was the fact that the overlookers, as a union body, refused to recognize the Council and did not send the representative to which they were entitled. A second major factor might have been that the realistic conception of the Council acting as an information pool had not been clearly separated from the idealistic conception held by some operatives of the Council acting in an executive fashion in relation to production matters. It seems highly likely that however clear the functions of the Council might have been to its members, they were not properly understood by most of the operatives in the mills.

THE PATTERN OF COMMUNICATION

Communication is one, perhaps the major, part of interaction, and is the important link in the building up of a relationship. This, then, is the justification for submitting to analysis communication as an activity.

There is no doubt that in both mills the immediate supervisor is the vital communication link for the operative. But beyond this, his function seems to vary as between weaving and non-weaving operatives. In the non-weaving case, he disseminates both information and instruction to his workpeople and spends some time in actually supervising work in progress.

In the case of the weaving supervisor, however, we noted a difference in emphasis which was very clear at Debenham; it was noticeable in the automatic weaving section at Radbourne but not so clearly as at Debenham. This difference was that the supervisor's prime task seemed to be to act as focal point for communication and action in relation to the continuous demands of automatic production. Purely supervising activity was of less importance, since in many respects the automatic looms themselves determined the rate and continuity of work. This would probably have emerged more clearly at Radbourne, even in these first stages of the innovation, had the supervisory and managerial functions of the shift managers not been confused.

At Radbourne, the fact that the present supervisory situation was in a state of flux, as the supervisors faced the demands of innovation, was demonstrated by the difficulties in working, about which the operatives complained. But it was of interest that the direction in which communication practice seemed to be travelling was that of the Debenham model, although there was little explicit transfer of the features of the one mill to the other.

At the beginning of this chapter we outlined three areas of information that differed in their degree of centrality to the working task, and we found here, too, interesting differences between automatic weaving departments and non-weaving departments that were true of both mills. One was that the peripheral areas of information were probably more important to the weavers and overlookers than to others. It is difficult to draw out exactly the reasons why this should be so, but our strong impression was that the automatic method brought with it a wider outlook on the part of the people actually doing the work. For the others, the little world of their particular department still represented the totality of their desire for information – even at Debenham. In so far as we have used a 'field' argument in our analysis of the mill – i.e. changes in one part of the system affect other parts of the system – this evidence gainsays that particular assumption. But it does seem that in many respects the various tasks preparatory to weaving had led traditionally to the creation of relatively closed fields of activity of warping, pirning, and so on. We have seen in Chapter Five that the automatic techniques were breaking down this tradition of isolation within the management group, but it did not yet appear true of the rest of the mill – even at Debenham, where physical boundaries were minimal.

The different perception of the activities of the mill as between operatives and management group does not appear to present problems at this stage of technology in the weaving industry. But future events may well alter this.

REFERENCES

MILLER, G. A. (1951). *Language and communication.* New York: McGraw-Hill.

Works not specifically referred to in the text

BLAU, P. M. & SCOTT, W. R. (1962). *Formal organizations.* San Francisco: Chandler.

BURNS, T. & STALKER, G. M. (1961). *The management of innovation.* London: Tavistock Publications; Chicago: Quadrangle Books.

TEN

Conclusions

I. GENERAL

The preceding chapters of this book have described our observations and analysis of a factory during a period of technological innovation. In this chapter we will draw together some of the main findings that we believe to be of psychological and practical importance.

During the period of our study a textile company that already had one totally automatic weaving mill introduced this type of technology in a section of another of its mills by converting one large non-automatic weaving department into an automatic one. However, the rest of the mill was not involved and the automatic section began operating literally in the midst of unaltered non-automatic sections. The innovation was thus only partial to the total technology, and for this very reason an interesting one to observe. Most technological innovation in industry is of just this partial character, and it is also usually introduced into an already established technology. The more spectacular cases of innovation, where the whole technology is changed in one step, are rare, and, as we shall suggest, may be psychologically and sociologically easier to achieve than a partial innovation.

Furthermore, automation in the textile industry is still in its infancy in this country. Burns (1956) has commented that while the ability of industrial companies to assimilate technical change successfully is of obvious importance to the companies themselves, this success also has great importance for the whole national society. And weaving has been for many years a very essential part of our economy. The whole nation has interest in the survival and development of this industry, and these are intimately related to the type of innovation that is the subject of this book. It is hoped, therefore, that some direct and useful information has been obtained in our study of the pioneering efforts of one company.

The aim of our study was to observe the interplay of social forces that arise in a factory (or a socio-technical system, as we have described it), when its pattern of productive behaviour is disturbed by the introduction of a new technical means of production. We have tried throughout to justify this description of a factory system in which there is an interplay of social and technical factors. On the one hand, the machines are described always in terms of the persons who use them, and, on the other hand, the individuals and groups in the factory are considered as existing together in this situation only because of the total task of production which in sum is defined by the very machines themselves. To study the psychological and social factors by themselves, in isolation from the machines and technical processes, would, we believe, have been an unfruitful and basically unsound procedure. At this stage in our knowledge of the behaviour of man and groups in this 'society in miniature', the factory, it is not possible to predict or formulate the interplay of these social and technical factors, which would make such separate, and more rapid, study possible and meaningful.

Because this view of the relation between technical and social factors was a fundamental assumption of our study, more than half our time (amounting to several hundred hours) was spent simply being in the midst of the machines, the men, and the cloth of the mill. When we did interview individuals, it was against the familiar background of their place in the technology, their behaviour in relation to their machines and the product, and the pattern of social contacts with others in the mill that they experienced while on their jobs.

What conclusions can be drawn from such a case study? An hypothesis is a statement of relationship between two or more concepts or elements that are meaningful and important aspects of the subject under consideration. It is part of the process of science to discover these meaningful elements and concepts in an area of knowledge, just as it is a later step in the scientific process to subject any hypothesis about these concepts to the test of practice or experiment.

It is our belief that the uncovering of meaningful elements in a situation, and perhaps the proposal of hypotheses about their relation, are the sorts of finding that can come from case studies such as ours. Hence part of this conclusion will be a summary of the factors that appeared to us to be important in the innovation situation. In addition, the observation of the day-to-day situations in the mill provided experience and some conclusions which may be of practical importance to those in industry whose lot it is to initiate and live through the ever-recurring process of technological innovation. In other textile companies carrying out similar innovations of automatic weaving, the experience of this company in solving its problems may be directly applicable. Furthermore, if our suggested concepts are valid for industrial social systems in general, the analysis of this innovation should provide useful suggestions for meeting problems of technological innovation in industry at large.

Finally, in this chapter we shall suggest the direction that further research might take, in order to increase further our understanding of man's behaviour in the industrial milieu which is so much part of our national society.

Summary of the Present Study

Radbourne Mill prior to this period had been predominantly non-automatic, and the innovation meant a definite replacement of this type of technology by the automatic one. A detailed description of the innovation has been given in Chapter Four. It is important to emphasize here that the actual time taken to complete the installation of all the new looms was a period of almost five months, but for at least comparable periods, both before and after, actual changes in organization and personnel, related to the innovation, were occurring.

These organizational changes were initially directly related to the new section of automatic looms itself. Operatives were selected, trained, and placed in charge of the new looms and supervisors for each operating shift were appointed. Then, as the innovation proceeded and the demands of automatic operation were felt, other changes in the organization of the remainder of the mill were made.

Management and the Innovation

The demands of the automatic technology arose in the main from the increased speed and continuity of operation of the new looms, the need for better quality warps and wefts, and the high initial capital cost of the equipment. The first and last of these features meant that downtime of the looms had to be minimized by ensuring an adequate supply of warp and weft to keep them running as continuously as possible.

The effect of these demands was felt, outside the automatic section, largely at the management level, although there was also some evidence of effects among operatives in other departments when the behaviour and attitude of their supervisors altered.

The increased need for planning the distribution of work in the various departments of the mill and for timing supplies in relation to departmental requirements was met by new appointments within the structure of the management group. These new roles all had a coordinating character, and in this way the social structure was brought more into step with the new technology. However, apart from the assistant manager, the occupants of the new roles were personnel who had held other clearly defined roles in the structure prior to the innovation. As the new technology began to operate, it was soon evident that these formal structural changes did not automatically imply that the new roles would be perceived accurately and that new patterns of behaviour in relation to these roles would rapidly appear. In fact, the attitudes and modes of behaviour towards the personalities in the new roles tended to persist as they had been in the non-automatic culture. This was evident from the interaction pattern of the management group. Apart from the assistant manager and the manager, the pattern was of a weaving management and a preparatory management group. The persistence of the traditional cleavage of the management group was also seen in the new meetings – formal and informal – which were set up to meet the need for increased communication in the organizing of the new technology. The membership of the sub-groups that appeared early in the innovation period closely followed this cleavage, but the last of the groups that emerged had members from both the traditional sections and indicated that new values and a new culture were beginning to emerge.

The accepted pattern of relationships between the management members in the non-automatic era had given the departmental heads a large measure of autonomy. Each was left to develop his own section of the mill with that section's specific task and interests as his pre-eminent aim and source of values. In contrast, the automatic technology emphasized the total task and the interests of the automatic looms as the focus of the whole mill, thus implying the development of a pattern of relationships which emphasized the management group as the organizing unit, rather than a set of independent organizers. This involved considerable re-assessment of the meaning of status and responsibility for the various management roles. Embedded as these were in the traditional culture of the mill, and interwoven with the personalities of the individuals, it was not surprising that the mere appearance of new positions in the social structure did not solve all the needs of the automatic technology. Nevertheless, the importance of such formal expression was paramount, since without it the development of the new relationships would have been still more confused.

The effect of the new roles in the social structure is most clearly seen in the case of the assistant manager. In this appointment, unlike the other changes in the structure, the occupant had not previously held another role in the mill. Unencumbered by the persistence of old relationships, personality identifications, and behavioural patterns, this person was able to initiate his own pattern of relationships and behaviour.

Furthermore, these were perceived and accepted in terms realistic to their function by the other personnel in the mill. In many ways, the appointment of the assistant manager was a major contribution to the successful innovation. This was due in large part to his specialized knowledge of automatic weaving; but also to a not inconsiderable extent to his ability to take advantage of his new role in the structure and establish appropriate behaviour patterns.

The formal changes in the structure in the management group were all aimed at the integration of organizational function. Although this integration brought the management group more into line with the demands of automatic weaving, it did mean that the operatives, still sharply differentiated from each other, found themselves with less and less contact with management. The traditional channel of communication upwards from operative to management was through the immediate supervisors. Both in the automatic loom section and in the warping department, there were difficulties when this traditional channel was weakened by the new appointments. No comparable formal reorganization of the operating groups appeared, and no new formal channels of communication for them were established. In fact, difficulties were met in both cases by introducing new supervisory roles into the management or junior management structure. However, among the automatic weavers and overlookers, it was increasingly observed that informal direct approaches for information were made by the operatives to the higher management. Some such new patterns of communication may in time be established, and these may equally well offset the effect of the changed management roles and remove the persistent need for ever-present supervision.

The Operatives and the Innovation

The two main groups of operatives directly affected by the innovation were the weavers and the overlookers who took charge of the new looms. Most had had at least some previous working experience of non-automatic looms but, in both groups, the new looms meant considerable changes in function and responsibility. For the weaver, the automatic looms meant that his direct manual contribution to the cloth in a loom was much reduced, whereas indirect physical contributions and supervisory functions were much increased. For the overlooker, the new looms were much more complex and more precisely designed than the old, and his behaviour towards them had to become much more standardized. As well as these quite distinct changes in the relation between these operatives and their automatic looms, there were other equally important changes for both, which related to the whole sett of looms, so that the sett, rather than the individual loom, became the functional unit for both types of operative.

The two groups of operatives did differ in their perception and attitudes concerning these loom-centred and sett-centred aspects of the innovation. The weavers became aware of the sett-centred features and as a group developed a favourable attitude towards the innovation more rapidly than did the overlookers. We suggest that this difference is related to the cohesiveness of the groups. Before the transition, the weavers were a group of low cohesiveness, while the overlookers had always been a highly cohesive group. The transition to the automatic section increased the cohesiveness of the transferred group of weavers but did not so affect the automatic overlookers. Among the weavers, increased cohesiveness indubitably predisposed the operatives to view the innovation with favour. Nevertheless, their perception of the innovation and their attitude to it were determined by frames of reference which stemmed from past experience in the old pattern of the mill. As their contact and experience of the new looms grew, the frames of reference of most of the operatives gradually changed from 'non-automatic' to 'automatic' – a change that was facilitated for the weavers by their increased cohesiveness. The loom-centred attributes of the automatic loom could be perceived and evaluated early in the innovation, since they fitted easily into the non-automatic frame of reference with its single-loom emphasis. A true sett-centred perception was achieved only when new frames of reference had emerged.

Despite the differences between the overlookers' and the weavers' groups in the way that they perceived the task-area of the new looms, individuals of necessity developed much closer ties. This was so because the tasks of the two workers became much more interdependent with the increase in the supervisory aspect of the weaver's task and the increase in the technical complexity of the overlooker's maintenance task. This interdependence was really novel and meant that a new basis for organization was possible – even if for the present unused.

Operative Groups and Innovation

Inter-group comparisons, as well as inter-group effects, play an important part in perception and attitude formation. The partial character of the innovation meant that non-automatic weaving sections continued to co-exist with the new automatic section. In fact there was not even a physical barrier of separation between the two sections. This meant that operatives in the new section were still surrounded by their old behavioural practices and culture. In addition to this, the new features of their automatic roles were only slowly recognized and given formal expression in the social structure. This encouraged the persistence of the non-automatic frames of reference, which then proved somewhat resistant to attempts by the management to introduce specific cultural modes relevant to the needs of automatic weaving.

This comparison with existing groups and their modes of behaviour also had a marked effect on the assimilation of the new roles by the operatives. The traditional status factors relating to job function, which still applied to the non-automatic groups, persisted in the automatic section long after the transition. These were unwittingly maintained by the management, as well as by the operatives, particularly in the modes of reward and sanction that existed for job performance. The integration of the functions of the automatic weavers and overlookers and of the functions of shift-opposites was not directly acknowledged by any new modes of payment. Furthermore, the traditional conception of 'skill', as it was understood by the operatives and management in non-automatic weaving, was found to be markedly inappropriate to the new automatic roles with their increased supervision, their interdependence on other roles, and their new emphasis on the sett of looms. Thus we found that a favourable attitude to the new looms did not always mean assimilation of all the features of the new role.

The emergence of new operative relationships such as those we have just mentioned, was a most interesting feature of the innovation. These arose directly from the interdependence of functions in automatic weaving. For the relations between shift-opposites, no established pattern was available, since shift work was not the common pattern prior to the innovation. Nevertheless, quite standard behavioural patterns did develop for these inter-shift relations.

These patterns remained without formal expression, except for one aspect of the relation between the pair of overlookers on opposite shifts. The overlookers as a group in the automatic section formally established a system of sharing the bonuses of shift-opposites. It is interesting to note that even in the highly cohesive overall group of overlookers this practice was only uniform for shift-opposites in the automatic section.

The new inter-group relationships were hindered by the traditional separation of the groups of weavers and overlookers in the culture of the mill. Nevertheless, the interdependence of a weaver and his overlooker became increasingly apparent in the automatic section. The increased overlap in the new roles and the geographical dilution of the weavers throughout the section, both emphasized these new relationships. The weaver or the overlooker was increasingly the only frequent face-to-face contact for the other while working within the sett of looms. New patterns for the relationship within this unitary work group did begin to emerge informally, as did new perceptions of status and the meaning of 'skill' in these new roles.

Training for Innovation

The training programmes for the automatic operatives fitted into this analysis of the non-automatic frame of reference. Thus the training was almost exclusively aimed at the loom-centred features of the new roles. This was very successfully handled; but the lack of training for the sett-centred features of the roles may have been responsible for the failure among the operatives to assimilate such features of their new role as systematic supervision.

One particular aspect of the management's handling of the training programme was of considerable importance to its success. This was the organization of the overlooker's training programme in such a way that it was consistent with the norms of this group and with their traditional relationship to management. This appreciation by the managers of the part such a cohesive group can play in training and reorganizing themselves was a recurrent feature of the manager–overlooker relationship throughout the innovation.

Innovation and Shift Work

The introduction of the automatic looms meant that a shift system of working was finally established as the working mode for the mill. Previously, only a small minority had been on shifts. Shift work was considered atypical and unnatural, and the mere establishment of a shift system in the major production sections of the mill did not automatically produce favourable attitudes towards it. Shift work had wide ramifications for individuals, involving areas of their lives that had hitherto appeared to be remote from working life. This meant that any assessment of the shift system of working was influenced by persons and features of life outside the four walls of the factory. In turn, since the shift system was inextricably related to the innovation, these external factors also influenced the operatives' attitudes to the new looms, their own work roles, and the whole automatic method of production. This was particularly true for those operatives for whom contact with the new looms and experience of shift work were concurrent. For several of these operatives, the external features of shift work, particularly its interference with their valued pattern of family relationships, were so dominant that the actual technical changes within the factory were largely disregarded.

Communicating Innovation

Each of these new relationships at all levels of the mill's structure had as a dynamic feature the communication of information. The automatic technology called for an enhanced flow of information between the various roles in the structure. The informal relationships that slowly emerged and the formal changes in the structure were all attempts to provide this communication. Sometimes they were successful, as in the case of the new management roles; but these, in turn, we have seen, interfered with the traditional operative-to-management channel for communication.

In all these relationships, the efficiency of communication depends on a common understanding of the words that convey the information. This understanding is fostered by a common experience of related behaviour, but often during the innovation period communication broke down because behaviour and information did not match. Only as new frames of reference emerged did common meaning and understanding begin to appear. Words like 'number of looms', 'sett', 'work-load', 'skill', 'supervision', and many others had to acquire new meanings in the minds of many individuals before efficient communication was achieved. Each person inevitably faced the new situations with his own experience and background up to that point. That these did not always produce a common view is not surprising. Rather, we should report the extent to which new common views did rapidly emerge, and the considerable contribution of the assistant manager in this respect. His ability to recognize and accept frames of reference other than his own was an essential part of this contribution to the efficiency of communications in which he was involved.

The general attitude of the management was to facilitate communication about the main features of the innovation, particularly through the Works Council. Although this information did not always reach the main body of the operatives officially, the general atmosphere of free communication persisted, since the mill was of such a size that face-to-face contacts at all levels could easily be made. The availability of face-to-face contacts and the trust they inspired were also important elements in the success of the innovation.

II. IMPLICATIONS FOR MANAGEMENT

At the beginning of this book we mentioned the textile company's traditional concern for the welfare of its employees and for the existence of good human relations, both formally between the management and the unions and informally between individuals. There is no reason to suggest that the innovation of automatic looms will mean anything other than the continuance of this pattern. The innovation with its many, often unforeseen, problems has been made without serious technical or social breakdown. Furthermore, it has been handled to a large extent by people who were in the mill prior to the innovation, and not by outsiders.

When innovation is made in this way, a company faces many more problems than when the new technology is introduced and run by new, experienced personnel. For an innovation made within an existing framework calls for changes not only in technology, but also in roles and responsibilities, attitudes and skills. In the former case, the problems of training are avoided and the effects of changes in structure and culture on attitudes and behaviour are not confused by the difficulty of perceiving the same faces and personalities in new roles. Thus we believe the company we have studied can make a real contribution through this record of its successful innovation. The measure of this success is reflected in the production figures from the automatic section at the close of our study. These were considerably in advance of those that had been achieved at Debenham Mill at a comparable time after the innovation there.

It is against this background of *successful* innovation and *absence* of social breakdown that our analysis of the innovation should be read. It has often pointed up the difficulties and discrepancies between ideal requirements and actual behaviour, but if these arise during innovation by an experienced and successful company there is every reason to suppose that they will arise in the innovation attempts of many other companies.

Much of what has been described and at times laboriously analysed is, for this company, past history and such a part of its everyday life as to appear almost trivial. However, some of the human problems that occurred unexpectedly from time to time during the study, can, if left unresolved, cause considerable losses in production efficiency and much personal unhappiness and dissatisfaction. We shall now, therefore, describe some practical points for the management of innovation which arose directly from the experience of this company.

The innovation of 112 automatic looms was much more than the addition of so many machines to a technology which already had a great many machines of a different type. It had ramifications over a long period of time both in the organization, and in the behaviour not only of the individuals in the automatic section but throughout the whole mill. Partial innovation of the sort that we witnessed also brings difficulties rather different from those arising in total innovation. Therefore the impact of technological innovation must be seen as affecting technical and organizational areas, together with the attitudes and behaviour of individuals and groups directly and indirectly concerned with the physical elements of the innovation.

Technical problems associated with innovation can be very real, particularly if the company is trying to adapt to its own use machines and equipment which may have been designed and tested in another context. However, what appear to be solely technical problems are quite often greatly complicated by individual and social features. Answers for some of the technical problems that arose in the operation of the automatic looms were already available, but the very words in which these answers were expressed created further problems. The same word could take on different meanings for the experienced manager and for the operative. The latter would draw his meaning from the experience and background of a non-automatic culture. As the culture changed, so the technical problem disappeared. Management must always be aware of this discrepancy in meaning as a source of technical difficulty. Words like 'supervision', 'adjustment', 'systematic method of working', 'responsibility', 'skill', and many others are heavily dependent for their meaning on the traditions and background of a company.

Changes in Social Structure

Organizationally, it is important to have a formal social structure which fits the new work tasks: new tasks mean new roles, and these are perceived and understood more easily if they are expressed in terms of the formal structure of the factory. In changing the social structure it is not sufficient to change a name. A position in the structure carries with it a host of factors such as wages, permitted relationships, responsibilities, and formal methods of reward and sanction. Consideration of these is also important if the structure or type of organization is to support the personnel in accepting and fulfilling their new tasks. Whenever the function of an individual is changed, his training must cover all the features of his new task. It may be that some tasks will change so radically that new types of personnel are required. A careful analysis of all the features of the new task should indicate the kinds of individual ability that are required. For example, in the present case it was not possible to train an individual adequately for a task involving *many* machines by familiarizing him solely with *one* of the machines. This sort of training calls for imaginative methods and a constant effort by management to avoid using only those methods which they have always used, and which have worked for earlier technologies. Similarly, traditional criteria for the selection of operatives, such as those of age or sex, may cease to be so important in the new setting. Furthermore, a highly suitable and well-trained operative may not perform adequately in his new task if he finds that the procedures of reward and sanction for the task are as they were in his old role. These procedures need to fit the features of the new task, the special training, and his new position in the organization.

When the changes in task are as radical as in the present case, the meaning of skill requires special attention. A factory where skill has been traditionally associated with manual operations of greater or lesser difficulty cannot suddenly and easily identify non-manual operations as skilful. Skill is an important status factor in industry, and the changes in task that accompany automatic innovation can easily disturb its stabilizing effect on the structure. But the formal definition of these new 'skills' by training and reward can aid perception and acceptance of their meaning.

Forming New Relationships

Rapid technological change is characterized by the emergence of new relationships throughout the social system. Recognition of these relationships is an important part of the management of innovation particularly as they may involve people who did not previously need to meet in direct relationship with each other. A formal definition at least of the new relationship may be achieved by bringing the individuals together. Individuals who share machines on different shifts need to be as much (often more) in relationship as people working together on adjacent machines. If this relationship between shift-opposites is important in the technology, then opportunity for these people to be physically together may need to be provided. This implies that an overlap between shifts should be regarded as part of the official task and rewarded accordingly.

On the basis of this study, we can say that management may well be advised to allow the personnel in such relationships to choose themselves. All relationships, even the most formal ones arising from the work task, include the evaluation of the individuals concerned by each other according to liking or dislike. Allowing for this factor by encouraging mutual selection may be a better means of establishing a good relationship than by defining its members always from outside. This appears to be true particularly when the relationships will be between members of a group of similar individuals performing the same task, and applies most strongly where cohesiveness of the group is thought to be desirable (see Trist *et al.*, 1963).

One other type of relationship which is worthy of management consideration is that between the people whose cooperation leads to the total completion of the production task. This work group is often not clearly defined in the organization of a factory (see Selznick, 1957). It consists of a vertical section of the social structure and its members perform the different tasks that make up the whole production process. The breakdown of jobs and the specialization that characterize modern industry often mean that this group of people is rarely together, either in time or place, in the factory. The re-integration of these individuals into a real group can lead to enhanced efficiency and genuine satisfaction.

Automatic technology in the present case meant greater interdependence for individuals belonging to different groups in the mill and also an increase in the differentiation of the total task, particularly at the unskilled level. It also meant that the usual type of peer work group – namely of operatives performing the same task – was less easily formed because of the physical distances between such operatives.

Of these features – increased interdependence, greater task differentiation, and less opportunity for peer grouping – the first and last increased the reality of the 'work' group, at least for the more skilled operatives. And this appeared without any formal recognition by the management. Greater task differentiation is very different in its effect, and it is not clear how it operated in this case. In the abstract case too great a differentiation, notably at the unskilled level, results in the body of such workers having only slight contact with the skilled group. But it is often possible to reintegrate the unskilled tasks by making them all the responsibility of a single individual, who will then be confined to a much smaller working domain and will therefore experience increased contact with a smaller number of work groups of skilled operatives.

The above descriptions of the work group have been made in terms of operatives only, but the full work group will include also the appropriate management members of the vertical section. How to give reality to this whole group is not clear even in an automatic technology. Nevertheless, the rewards of so doing may be great, since it is, in effect, through the efforts of this whole group that production is achieved. It is only by actual recognition of this interdependence that the traditional divisions between labour and management with their conflicting interests can be properly assessed and bridged. Until the present, the organization of industry at both management and operative level has increasingly tended to differentiate roles and place them in a hierarchy. Such a structure inevitably establishes barriers in a vertical direction, while stratifying relationships, beliefs, patterns of behaviour and self-interests in the horizontal direction. The pursuit of goals associated with these strata then becomes one of the main sources of satisfaction for the members of the organization. The discrepancy between this type of structure and the relationships directly demanded by the task of production is obvious. Awareness of the importance of all these relationships is essential for those in industry who are concerned about job satisfaction and productivity.

Resistance to Change

Whenever 'resistance' to technological change appears, it can be taken as indicative that new relationships are not being successfully achieved or, rather, that they are not leading to improved satisfaction. Individuals and groups in a factory seek their own interests in the form of wages, enhanced respect, better working conditions, greater job satisfaction, and in many other ways. If 'resistance' can be seen as a positive seeking after these self-interests rather than a negative blocking of progress, the management of innovation will be simpler. It was striking to note that the group in the mill whose status and corporateness increased considerably after the innovation rapidly achieved a favourable attitude towards it. Particular attention needs to be given to the way highly cohesive groups, whose cohesiveness is already one probable source of satisfaction, are to be affected by the innovation.

Communication is the link that builds relationships, and effective communication is an essential part of the management of innovation. Again, communication is not merely a matter of words but must be supported by appropriate social organization, new modes of behaviour, and attitudes which fit the new situation. The communication of information about innovation must be 'translated' in such a way that it is understood in the same sense by people whose traditions and previous experiences have been different. No single set of words can achieve this end. Nor will these differences in understanding become apparent unless the different groups are brought face to face with each other. The present study has much to contribute here in its description of the management group. The confrontation of the previously autonomous members of this group with a task which demanded interdependence and rapid communication, called for changes in both group structure and normal channels of interaction. New groupings, both formal and informal, emerged and gradually new and more appropriate channels of communication for the new situation were developed.

Communication down the social structure was important for those directly involved in the change and subsequent operation. The importance of negative information about the change was also apparent. When innovation occurs in part of a factory, uncertainty can arise in other parts, and this can best be met by adequate information and reassurance to those who will not be affected by the change. During innovation the importance of communication up the social structure is also enhanced. Only in this way can the management keep in close touch with the progress of the innovation at its vital points, namely where it affects the operative whose task is the immediate operation of the new machine.

Most of these points will be seen to be merely an extension of the problems that constantly confront management; but it is during innovation that their incidence can increase to such an extent that breakdown is threatened.

III. SOCIAL SCIENCE CONCEPTS AND THEORIES

Our final concern is to relate our broad conclusions to more general scientific concepts and theories drawn from sociology and psychology. Our material illumines some concepts – particularly the explicitly social psychological – and leaves others unrefined. This was inevitable from the roles which we developed as research workers and also our decision to deal with those data which contributed most to our understanding of this particular type of socio-technical operation.

Industrial Small Groups

We will start by considering the nature and types of the small face-to-face groups with which we were much concerned throughout the study. Most of the groups that have been the subject of social psychology are *peer groups.* A peer group seems to have certain properties that differentiate it from other sorts of group, and these are:

1. A common property of age, skill, responsibility, or the like, which is possessed by each of the individual members.
2. Election to the group rests both with the individual requesting membership voluntarily and also with the members jointly granting or withholding membership.

3. Selection of leaders is restricted to the group members, whether such leaders are implicitly or explicitly appointed.
4. Activities undertaken by the group are often but not always determined by the group members.

In recent years, a number of distinctive behavioural features characterizing such groups have been found. For example, the tools of sociometry and interaction mapping have revealed the close relationship between the likes and dislikes within the group and its pattern of interaction behaviour. Festinger and others (Festinger, 1954; Festinger *et al.*, 1954) have devised means of measuring the cohesiveness of peer groups and shown that the degree of cohesiveness is closely related to the uniformity of behaviour patterns. The classic experiments of White & Lippitt (1954) on leadership atmosphere in boys' clubs can also be analysed in terms of cohesiveness when a relation between cohesiveness and the friction among the group members becomes apparent.

If we now consider the groups that appear in an industrial social system such as a factory or industrial company, it is possible to discern two distinct types of possible grouping. The first bears a close resemblance to the general peer groups we have just described. Thus there are a number of managers, a number of line foremen, a number of mechanics, engineers, cleaners, machine operators, etc., who occupy the various roles in the social structure. Each member of these groups has in common the social and technical status factors of that role and exhibits the same pattern of activity – an element in the total behaviour of the social system. The 'similarity' property of the peer group holds in all these cases and so do some of the other properties outlined above, but not all of them. Membership of the role is often by free choice although which specific role group is decided by a person outside the group. Powers of expulsion from the group are officially vested outside the group but can nevertheless be unofficially exercised by the group in some cases. The formal group leader is appointed from outside the group and is not a group member, but the groups informally (and formally in the case of union stewards and representatives) do elect their own leaders. The task or activity of the group is not at first sight determined by the group, e.g. their behaviour in relation to the total production is defined from outside the group, but such groups frequently select for themselves other tasks which serve to maximize some or all of the satisfactions associated with the performance of primary technical operations.

It is clear that, in terms of these properties, this type of industrial group is very similar to the peer groups defined above. In some circumstances, however, industrial groups operate in ways that distinguish them from these general peer groups, as we shall call them.

Studies of groups in industrial systems have been almost exclusively concerned with industrial peer or role groups of this type and both aspects of their existence have attracted attention. Homans (1953) has considered groups of girls in an office simply in terms of their peer group behaviour. His findings agree entirely with the pattern of liking and interaction referred to above. Again, the formation of in-groups, which is mentioned repeatedly in the literature on free peer groups, was found in the very early Hawthorne studies of the bank wiring group (Roethlisberger & Dickson, 1939). On the other hand, the non-free aspect of the life of these groups has been the focus of such studies as those of Katz *et al.* (1951a, 1951b) in the U.S.A. and of Argyle, Gardner, and Cioffi (1958) in Britain. The investigators have been concerned with the production task of the group members. Productivity has been measured, along with such other variables as type of supervision, group morale, and job satisfaction, and relationships between them have been explored.

The results have been in many ways confusing and apparently paradoxical. Thus, though some relationship does appear to exist between type of supervision and productivity, no clear link has been found between group morale or job satisfaction and productivity. This latter generalization has also been found in a number of case studies including the present one and that of Homans (1953), who without further comment remarks, 'contrary to the theory that talking interfered with work, the figures show no correlation between output and frequency of free choice interaction.'

Against this background of our present knowledge about the first type of industrial group, we can now define the second discernible type.

This group, which we call a *work group*, consists of those people in a social system whose direct cooperation completes the total productive task of the system but whose sub-tasks are different. These groups are thus defined by the total task, that is to say, by a property external to the members, in contrast to the internal definition of the first type of group. When the other properties of these work groups are considered, they are found to differ at all points from the general or industrial peer groups. Membership is defined from outside the group and depends very heavily on such features of the system as the geography and the degree of differentiation of total production. The members occupy different roles in the social structure, and the group forms a vertical section of it. As this vertical section is ascended, each member has some responsibility for, and leadership of, those in roles on the lower levels. Expulsion of lower members of the group is one of these responsibilities at some of the levels. Further, it is rare in most industrial systems, for these groups to have psychological reality in the sense of being together as a group in space or time.

Now that we have completed our definition of the two types of group in industrial social systems, we wish to suggest the following explanation of the confusing and paradoxical findings referred to above. There appears to be a confusion between the nature of the group and its real task. Groups of the first kind do not exist solely for the task of production. These groups have other self-appointed tasks which may in fact conflict with maximum output. Nor is the cooperation between the members of an industrial peer group a sole determinant of production. The production of this group depends on the behaviour of many people in the system who occupy roles other than the one which allows membership of the group. Thus, though we may expect a correlation between the morale of these groups and their achievement of such in-group tasks as the improvement of their status in the system, we need not expect a simple correlation between morale and productivity. This is precisely what Homans's case study indicates.

On the other hand, if we could identify and give reality to the work groups in the system, we would expect to find relations between their productivity and appropriately defined properties of these group. Two other major studies bear this out. The earlier one by Rice (1958), to which we have referred, and the more recent English mining study by Trist *et al.* (1963). In both these case studies, where work groups were given reality, both geographically and continuously in time, and formally recognized in a new social structure, productivity was found to be directly affected. In our own study, the work group as such did begin to emerge in part, but not in entirety except for the Debenham management. Wherever it did at Radbourne it appeared to be significantly related to productivity in a way that the other formal groupings in the structure were not. For example, the management group underwent many reorganizations, all of which were related to the total task of production. The record of their interaction behaviour broadly followed their functional relationships with each other. The shift-opposites – weavers and overlookers – constituted sub-work groups in which the type of relationship had a direct influence on the productivity from the common machines. Finally, the weavers and overlookers who shared common machines began to perceive themselves as a group, and this was of prime importance to their output.

It appears to us that the real nature of small worker–management groups in industrial social systems requires much closer attention. Even when the groups are defined, the problem of identifying and assessing their important features will remain. The final achievement of harmony between the shift-opposites at both Debenham and Radbourne appears to have depended on the groupings being defined by some free-choice procedure.

Cohesiveness in Industrial Small Groups

For the 'role' or 'peer' type of industrial group we have found it possible to redefine in some measure the concept of cohesiveness so that it becomes useful for the analysis of behaviour.

The cohesiveness of an industrial 'role' group is indicated by,

(*a*) the extent of union membership in the group;

(*b*) the existence of group-defined patterns of behaviour (for examples in this study, the enforcement of a starting time of work regardless of an arrival time, a group approach to the management about wage issues, etc.) and;

(*c*) the existence of common frames of reference and uniform attitudes as they appeared in behaviour and opinions. Furthermore, the increase or decrease of similarity of membership in such properties as age, sex, etc., are likely (but not essential) prerequisites for gain or loss of cohesiveness.

With these rather tentative definitions of this cohesiveness, the following hypothesis can be proposed as a result of our study. *When the cohesiveness of a role group increases as a result of a technological innovation, the establishment of new frames of reference and favourable attitudes to the change is more rapid than where cohesiveness is constant or is decreased.*

Before this hypothesis can be tested generally it will be necessary to define cohesiveness further and also to explore much more fully the effect of intra- and extra-group procedures on the reorganization of social structures. What has already been said may suggest that external reorganization can be satisfactorily achieved for groups of low cohesiveness but internal reorganization may be better for highly cohesive groups.

Dilution of Groups

Another aspect of the cohesiveness of role groups that merits investigation is the effect of geographical dilution. Automatic innovation tends to dilute, or spread out in space, the operatives who occupy the same role in the system. The present study suggests that 'non-interactors' in the role group establish appropriate frames of reference and favourable attitudes more slowly than 'interactors', despite the fact that non-interaction is a more appropriate behaviour pattern for the new task. The non-interactors may be such because of their personality or because of the previous social history of the situation. If the latter is the case it would be of interest to know how, in such a situation, they can be more closely integrated into the group. One possibility would be the effect of formally and explicitly recognizing by some status reward the new features of the task, which might facilitate acceptance by the group of the non-interactor, since new conditions of membership would hold with which he could comply. Further, membership of the group might now be more keenly sought by the non-interactor since it would be explicitly related to his real behaviour, prestige, and needs.

Finally, it would be interesting to know the relation between dilution of the role group and the establishment of the reality of the work group.

Other Types of Group Relations

The final major area that transcends the peer and work groups is that of inter-departmental relations. Here we draw much closer to notions of leadership in the formal sense. For one of our most striking findings was the need to abandon the old modes of intergroup relations and substitute a cooperative frame of reference. Because of the broad sweep of the operation and the relative lack of existing linkages, 'leadership' in the sense of establishing goals, setting limits, and actuating wider loyalties became very necessary indeed. In this situation, the assistant manager played a vital role exercising technical and organizational leadership in day-to-day operation. It is quite likely that, had this man not been available, some other person would have carried out the leadership task. What was striking was the way in which the need for some such role became very apparent. This brings to mind the type of analysis made by Selznick (1957) in a wider context than the present one, but dealing with similar issues.

Finally, in this study we have seen something of the interplay of the technical and human factors as one enterprise is moulded into a significantly different one. We have observed and reported individual and group ways of dealing with these complex events of technical change. Detailed observation is an essential ingredient of future studies if social science is to help in the understanding of these aspects of human behaviour.

REFERENCES

ARGYLE, M., GARDNER, G. & CIOFFI, F. (1958). Supervisory methods related to productivity, absenteeism, and labour turnover. *Hum. Relat.*, vol. XI, pp. 23–40.

BURNS, T. (1956). *Impact Sci. Soc.*, vol. 7, pp. 147–65.

FESTINGER, L. (1954). Informal social communication, in Cartwright, D. & Zander, A. (eds.), *Group dynamics: research and theory*. Evanston, Ill.: Row, Peterson; London: Tavistock Publications.

FESTINGER, L., SCHACHTER, S. & BACK, K. (1954). The operation of group standards, in Cartwright, D. & Zander, A. (eds.) *Group dynamics*.

HOMANS, G. C. (1953). Status among clerical workers *Hum. Organization*, vol. 12, pp. 5–10.

KATZ, D. *et al.* (1951a). *Productivity, supervision and morale among railroad workers*. Ann Arbor: University of Michigan.

KATZ, D. *et al.* (1951b). *Productivity, supervision and morale in an office situation*. Ann Arbor: University of Michigan.

RICE, A. K. (1958). *Productivity and social organization: the Ahmedabad experiment*. London: Tavistock Publications.

ROETHLISBERGER, F. J. & DICKSON, W. J. (1939). *Management and the worker*. Cambridge, Mass.: Harvard University Press.

SELZNICK, P. (1957). *Leadership and administration.* Evanston, Ill.: Row, Peterson.

TRIST, E. L., HIGGIN, G. W., MURRAY, H. & POLLOCK, A. B. (1963). *Organizational choice.* London: Tavistock Publications.

WHITE, R. & LIPPITT, R. (1954). Leader behaviour and member reaction in three 'social climates', in Cartwright, D. & Zander, A. (eds.), *Group dynamics.*

General References

ARGYLE, M. (1957). *The scientific study of social behaviour.* London: Methuen.

ARGYLE, M. *et al.* (1957). The measurement of supervisory methods. *Hum. Relat.*, vol. X, pp. 295–313.

ARGYLE, M. *et al.* (1958). Supervisory methods related to productivity, absenteeism, and labour turnover. *Hum. Relat.*, vol. XI, pp. 23–40.

BAVELAS, A. in LEWIN, K. (1952). *Field theory in social science.* London: Tavistock Publications, pp. 230–1.

BLAU, P. M. (1955). *The dynamics of bureaucracy.* Chicago, Ill.: University of Chicago Press.

BURNS, T. (1954). The directions of activity and communications in a departmental executive group. *Hum. Relat.*, vol. VII, pp. 73–97.

BURNS, T. (1956). *Impact Sci. Soc.*, vol. 7, pp. 147–65.

CARTWRIGHT, D. & ZANDER, A. (eds.) (1954). *Group dynamics.* Evanston, Ill.: Row Peterson. London: Tavistock Publications.

CONRAD, R. & SIDDALL, J. G. (1953). An experimental study of pirn-winding. The effect of mixed setts on operative efficiency. *J. Textile Inst.*, vol. 44, pp. 215–22.

CURLE, A. (1949). A theoretical approach to action research. *Hum. Relat.*, vol. II, pp. 269–80.

EMERY, F. E. & MAREK, J. (1962). Some socio-technical aspects of automation. *Hum. Relat.*, vol. XV, pp. 17–26.

FESTINGER, L. (1954). Informal social communication, in Cartwright, D. & Zander, A. (eds.), *Group Dynamics.*

FESTINGER, L., SCHACHTER, S. & BACK, K. (1954). The operation of group standards, in Cartwright, D. & Zander, A. (eds.), *Group Dynamics*.

FRENCH, J. R. P. & COCH, L. (1948). Overcoming resistance to change. *Hum. Relat.*, vol. I, pp. 512–32.

FRENCH, J. R. P., ISRAEL, J. & ÅS, D. (1960). An experiment on participation in a Norwegian factory. *Hum. Relat.*, vol. XIII, pp. 3–20.

HOMANS, G. C. (1953). Status among clerical workers. *Hum. Organization*, vol. 12, pp. 5–10.

JACKSON, J. M. (1953). The effect of changing the leadership of small work groups. *Hum. Relat.*, vol. VI, pp. 25–44.

JAQUES, E. (1951). *The changing culture of a factory*. London: Tavistock Publications.

KATZ, D. *et al.* (1951a). *Productivity, supervision, and morale among railroad workers*. Ann Arbor, Michigan: University of Michigan.

KATZ, D. *et al.* (1951b). *Productivity, supervision and morale in an office situation*. Ann Arbor, Mich.: University of Michigan.

LEWIN, K. (1952). *Field theory in social science*. London: Tavistock Publications.

MILLER, G. A. (1951). *Language and communication*. New York: McGraw-Hill.

NEWCOMB, T. M. (1952). *Social psychology*. London: Tavistock Publications.

RICE, A. K. (1958). *Productivity and social organization: the Ahmedabad experiment*. London: Tavistock Publications.

ROETHLISBERGER, F. J. & DICKSON, W. J. (1939). *Management and the worker*. Cambridge, Mass.: Harvard University Press.

RONKEN, H. O. & LAWRENCE, P. R. (1952). *Administering Changes*. Boston, Mass.: Harvard Graduate School of Business Administration.

SELZNICK, P. (1957). *Leadership in administration*. Evanston, Ill:. Row, Peterson.

STOUFFER, S. (1949). *Studies in social psychology in World War II*. Princeton, N.J.: Princeton University Press.

TRIST, E. L. & BAMFORTH, K. W. (1951). Some social and psychological consequences of the Longwall method of coal-getting. *Hum. Relat.*, vol. IV, pp. 3–38.

TRIST, E. L., HIGGIN, G. W., MURRAY, H. & POLLOCK, A. B. (1963). *Organizational choice*. London: Tavistock Publications.

WHITE, R. & LIPPITT, R. (1954). Leader behaviour and member reaction in three 'social climates', in Cartwright, D. & Zander, A. (eds.), *Group Dynamics*.

WOODWARD, J. (1957). Control and communication – a management concept of cybernetics. *J. Inst. Prod. Engrs.*, vol. 36, pp. 539–48.

Works not specifically referred to in the text

BLAU, P. M. & SCOTT, W. R. (1962). *Formal organizations*. San Francisco: Chandler.

BURNS, T. & STALKER, G. M. (1961). *The management of innovation*. London: Tavistock Publications; Chicago: Quadrangle.

CROSSMAN, E. R. F. W. (1960). *Automation and skill*. London: H.M.S.O.

HERBST, P. G. (1962). *Autonomous group functioning*. London: Tavistock Publications.

JACOBSEN, H. G. & ROUCEK, J. S. (eds.) (1959). *Automation and society*. New York: Philosophical Library.

KING, S. D. M. (1960). *Vocational training in view of technological change*. Paris: European Productivity Agency, Project 418.

LEACH, E. R. (1954). *Political systems of highland Burma*. London: Bell.

MANN, F. C. & HOFFMAN, L. R. (1960). *Automation and the worker*. New York: Holt.

MAREK, J. (1962). Effects of automation in an actual-control work situation. London: Tavistock Institute of Human Relations. Document 669.

NADEL, S. F. (1956). *Theory of social structure.* London: Cohen & West.

POLITICAL AND ECONOMIC PLANNING (1957). *Three case studies in automation.* London: P.E.P.

WALKER, C. R. (1957). *Toward the automatic factory.* New Haven: Yale University Press.

Author Index

Ås, D., 10
Argyle, M., 3, 10, 36, 39, 172, 177, 226, 231
Back, K., 231
Bamforth, K. W., 7, 11, 39
Bavelas, A., 2, 10
Blau, P. M., 3, 10, 205
Burns, T., 84, 88, 205, 206, 231
Cartwright, D., 8, 10
Cioffi, F., 226, 231
Coch, L., 11
Conrad, R., 95, 126
Crossman, E. R. F. W., 126
Curle, A., 2, 10
Dickson, W. J., 2, 11, 226, 231
Emery, F. E., 82, 88
Festinger, L., 8, 225, 231
French, J. R. P., 2, 10, 11
Gardner, G., 226, 231
Herbst, P. G., 178
Higgin, G. W., 178, 232
Hoffman, L. R., 178
Homans, G. C., 3, 10, 150, 172, 177, 226, 231
Israel, J., 10
Jackson, J. M., 2, 10
Jacobsen, H. B., 11
Jaques, E., 3, 7, 10
Katz, D., 3, 10, 172, 177, 226, 231
Lawrence, P. R., 2, 11
Leach, E. R., 11
Lewin, K., 4, 10, 120, 126
Lippitt, R., 225, 232
Mann, F. C., 178
Marek, J., 82, 89, 126
Miller, G. A., 179, 205
Murray, H., 178, 232
Nadel, S. F., 11
Newcomb, T. M., 7, 11
Pollock, A. B., 178, 232
Rice, A. K., 155, 172, 177, 228, 231
Roethlisberger, F. J., 2, 11, 226, 231
Ronken, H. O., 2, 11
Roucek, J. S., 11
Schachter, S., 231
Scott, W. R., 205
Selznick, P., 220, 231, 232
Siddall, J. G., 95, 126
Stalker, G. M., 205
Stouffer, S., 120, 126
Trist, E. L., 7, 11, 32, 39, 178, 221, 228, 232
Walker, C. R., 178
White, R., 225, 232
Woodward, J., 87, 89
Zander, A., 8, 10

Subject Index

absenteeism, 49
acceptance of the research workers, 1 f.
'action research', 2
administrative structure of the mills, 18 f.
age criterion,
  management's reasons for, 133 f.
  and uncertainty among operatives, 135 f.
Amalgamated Engineering Union, 17
aptitude tests, not favoured, 130
artificial silk, introduction of, 13
Ashby (non-automatic foreman at Radbourne), 45, 54, 199
  attends daily meeting, 86
  as shift manager, 49, 72, 173
assistant manager, 71 ff., 209
  importance of, 210 f., 230 f.
  relation to shift managers, 81
  responsibility for carrying out programme, 78
  see also under Laycock
automatic looms, 97
  absence of from Radbourne school, 136 f.
  acceptance of, 102 f.
  and communications, 187
  downtime on, Radbourne and Debenham, 59, 78 ff.
  emphasize weavers as group, 117
  given priority in preparation departments, 74 f.
  introduction of,
    at Debenham, 16 f.
    at Radbourne, 40 ff., 57 ff., 66 *et passim*
    reasons for, 14 f.
  and introduction of shift work, 15, 16
  manual for, 97, 98
  noise from, 123
  operation of, at Debenham, 26
  and pattern of working of mill, 35, 38
  production costs of, 14
  removal of packets on, 93 f.
  technical features of, 90 f.
automatic spooling machines, 63
automatic weaving, pattern of, 152 ff.

back beaming warping machine, 49
base wage, in work study, 24 f.
Battersley Mill, monthly meeting at, 34
battery filling, 92 f., 94 f., 153
beam, 20
beamer, 55
Black, 200
  attends daily meeting, 85
  responsibilities of, extended, 63, 76 f.
bonus money, 24 f.
  commencement of, for new looms, 52
  and increase in size of setts, 57
  recommencement of, after move, 48
  and shift working, 164
borrowing of yarn, between mills, 38
Bowyer, 128 f.
Boyd (cloth room foreman),
  views of, on weavers' role, 108
break frequency,
  and lack of humidification, 46

case study, reasons for, 4
chief clerk, at Debenham, 31
cleaners, 148 f., 153
cloth room supervisor, position of in Debenham organization, 31
cohesiveness, 150
  and frames of reference, 8, 229
  in industrial small groups, 229
communication, 179 ff., 216 f.
  attitude of operatives to, 190 ff.
    preparatory departments, 190, 192 f.
    weavers, 193 ff.
  and changing technology, 197 f.
  at Debenham, 184 ff.
  Debenham and Radbourne compared, 195
  distortion of, 180
  formal and informal, 183 f.
  general discussion of, 179
  inter-departmental, 186, 187
  inter-shift, 170 f., 185, 187
  methods of, 183 f.
  non-personal, 180
  at Radbourne, 186 ff.
  and resistance to change, 223
  and supervisors, 198 ff., 203, 211
  types of information in, 181 ff.
  upwards, 198 ff.
  work study, 187 ff.
company directorate,
  attitude of to the project, 1 f.
  method of filling new roles, 127
company headquarters, 18
  ordering of yarn by, 37 f.
  specialist advisors from, 32
conclusions from study, 206 ff.
continuity of working, 35 f.
Crompton and Knowles looms, 14
culture of factory, 80, 86 f.
  definition of, 6 f., 33
  independent of changes in social structure, 200
  non-automatic, persistence of, 209 f.
  and work groups, 172
daily production meeting, at Debenham, 33 ff., 185
damage,
  definition of, 24
  rate,
    rise in, 48, 52
    and working through meals, 56, 58
  sanctions for, 112 ff.
  and shift working, 162 ff.
  tickets, 196 f.
Debenham Mill, 26 ff.
  administrative structure of, 18
  age uncertainty at, 136
  cloth handled at, 37
  communication pattern at, 184 ff.
    attitude of operatives to, 190 ff.
  contact with Radbourne, 32
  daily production meeting at, 33 ff., 185
  'damage ticket' system at, 196
  downtime, compared with Radbourne, 59, 78 ff.
  helps Radbourne in overlooker shortage, 54, 58, 142
  history of, 15 ff., 28 ff.
  humidification plant at, 16 f.
  introduction of automatic looms at, 16 f.
  layout of, 26 f.
  management subgroups at, 84
  and national industry, 19
  organization of, 28 ff., 38 f.
    compared with Radbourne, 67, 68
  pairing of overlookers at, 166
  pattern of working at, 32 ff., 38 f.
  satisfaction with information at, 195
  supervision by overlookers at, 119 f.
  training at, 128
  Works Council at, 184 f., 202
  as yardstick for evaluation of Radbourne, 26
definitions, 4 ff., 224 f.
Downham, 53, 165
downtime, 48 f., 59

downtime—*contd.*
during innovation period, 61, 63, 78 ff.

early starting, 119 f.
efficiency,
at Debenham, 29 f.
definition of, 24
at Radbourne,
during changeover, 48, 62
during innovation, 61 f.
through meal breaks, 58
of weavers, 109 ff.
Electrical Trades Union, 17
enterers, 21
entering, 20 f.
department, no immediate effect of innovation on, 74
Evans, 165

factors in work study, 25
factory, as social system, 5 ff.
factory managers, attitude of, to the project, 1 f.
family life, effect of shift work on, 173 ff.
fell forecast, 77 f.
fitters, 148 f., 153
fixed-rate payments, on introduction of automatic looms, 45
dissatisfaction with, 46
raising of, 47
foremen, competition among, for trainees, 129
foremen's meeting, 84 f., 88
announcement of change to, 44
frames of reference, 7 ff., 125
and changes, 10
and group cohesiveness, 8, 229
and views of weaver's role, 109

Gardner (automatic shed foreman at Radbourne), 45, 199
attends daily meeting, 86
as shift manager, 49, 72, 173
views of, on weaver's role, 107 f.
geographical dilution of groups, 229 f.
Glacier Metal Company, 3
group cohesiveness, description of, 8 f.
groups, 115 ff.
behaviour of,
definition of, 4
formal and informal, 8 f.
overlookers as, 121 ff.
under new conditions, 115 ff.
unions and, 116 ff.
weavers as, 117 ff.

Hambly, 165
Harding,
lends Knight to Radbourne, 54
position of, in Debenham organization, 31
Harrison, Miss, 60
Hawthorne experiment, 2, 226
heating, at Radbourne, absence of during introduction, 45 ff.
helper, 148 f.
home life, and shift working, 172 ff.
hooter campaign, 49 ff., 118, 119, 133
humidification plant,
at Debenham, 16 f.
at Radbourne, delay in introduction of, 45 ff.

Illingworth, 165
Indian Mill, work groups in, 155
individuals, behaviour of, definition of, 4
industrial company, as social system, 5 ff.
industrial small groups, 224 ff.
cohesiveness in, 229
dilution of, 229 f.
information on change,
negative, importance of, 134, 223 f.
small effect of on operatives' attitudes, 123 f.
types of, required 181 ff.
innovation,
and communication, 216 f., 223 f.

innovation—*contd.*
and management, 70 f., 209 f., 217
and operative groups, 100 ff., 213 ff.
and the operatives, 212 f.
partial, 206
inter-shift relations, 159 ff., 214
and breakdowns, 161
Irwin, 43, 99, 165, 168
moves to new looms, 51
and effect on old looms, 52
view of, among overlookers, 144

Jackson (programming supervisor), 77 f.
anomalous position of, 86
junior supervisors, 68

Kaye,
lent to Radbourne, 54
and training of overlookers, 142, 144
Knight,
lent to Radbourne, 54, 144
return of, to Debenham, 58
training by, 143
view of overlookers, 98 f.
knotters, 22, 187
knotting, 20, 21 f.
knotting mechanic, 153

Lavenham, 44, 124
Lawson,
attends daily meeting, 85
and the innovations, 73
Laycock, 54, 56, 68, 82, 109, 113, 144, 158
action of, during wage and humidification troubles, 46
and age criterion, 136
arranges help for overlookers from Debenham, 54
and co-ordination, 78
at daily meeting, 86
experience of, 72
and hooter campaign, 50
move of, to Radbourne, 31, 32, 43 f., 72 ff.
and pairing of overlookers, 165 ff.
and programming, 77
views of,
on departmental isolation, 78
on male weavers, 133
on overlookers, 98 f., 143
on overlooker-weaver relationship, 112
on weaver's role, 107
see also under assistant manager
load, *see* work-load
'loom-centred',
view of changes, 98, 101
view of training, 143
loom mechanic apprentices, numbers of, 17
loom mechanics, *see* overlookers
looms,
cleaning and oiling of, 93
description of, 19 f.
felled out, 23
number of, per operative, 59

male automatic weavers, comments of, on female, 132 f.
management (and management group),
and announcement of changes, 123
and 'automatic frame of reference', 109, 133
change in, at Radbourne, 43 f., 63, 67 ff.
contact within,
at production meeting, 33 ff., 82 ff.
outside production meeting, 36
expectation of, of factors of importance during innovation, 123 f.
and hooter campaign, 50 f., 63
implications of this study for, 217 ff.
and importance of overlooker training, 141 f.
and innovation, 209 ff.
and inter-shift disputes, 164
and maintenance of shift pairings, 170
new appointments among, 209

management—*contd.*
and new relationships, 221 f.
organization of,
at Debenham, 30 ff.
at Radbourne, 42, 43 f., 64, 67 ff., 87
and overlookers as 'group', 123
persistence of traditional cleavage in, 210
puts pressure on sizing department, 74
and quality of production, 87 f.
and quantity of production, 87
relation between formal and informal structure of, 82 ff., 88
and resistance to change, 125
selects men for training, 130
reasons for, 131 ff.
social structure of, 5
and shift managers, 82
and specialization policy, 93
status levels in, 83, 87
sub-groups within, 83 f., 88
views of,
on overlooker's role, 115
on weaver's role, 107 ff.
and weavers as 'group', 117
'weavers' and 'non-weavers' in, 88
and work study, 24
manager,
assistant, *see* Laycock
assistant to, 71
managers,
and company headquarters, 32
contact between, 32
function of, 18
views of, on labour relations, 33
man-made fibres, 12, 13
Manning, 44, 137
married men, effects of shift working on, 173 ff.
meal breaks,
and hooter campaign, 50
and running looms through, 56, 59
financial incentives to, 57 f.
and social groups, 150 f.
meetings, 33 ff.
daily,
at Debenham, 33 ff.
at Radbourne, 82, 85 f., 88
functional and structural, 86
weekly, 84 f.
Monks, 52
Moore (workshop engineer), 188
views of,
on overlooker-weaver relationship, 112
on weaver's role, 108, 109
morale, and productivity, 226 ff.
Morris,
and damaged cloth, 196
position of, in Debenham organization, 31 f.
multi-loom setts, 124

National Association of Power Loom Overlookers (N.A.P.L.O.), 17, 116
agreements with company, 122
and pairing, 166
refusal to operate under work-study scheme, 17
terms of, for apprentices, 137
new relationships,
formation of, 220 f.
and resistance to change, 223
noise, increase of, no unfavourable reaction to, 123
non-automatic department,
change of location of, 42, 45, 65
compared with automatic, 152 ff.
foreman of, moved temporarily to automatic, 63
shift opposites in, 169 f.
traditional pattern of, 148 ff.
'non-automatic' frames of reference, persistence of, 213 f.
non-automatic looms, as yardstick for assessment of innovation, 100 ff., 109
nylon quality, 53, 56 f., 58 f., 60, 65
sort factor for, 57

nylon quality—*contd.*
suitability of, for running through meal breaks, 56

Oakroyd, 52, 53, 58, 82
absence of, during humidification trouble, 46 f.
and age criterion, 136
announces changes, 44
announces delay in arrival of new looms, 47
attends daily meeting, 86
and management changes at Radbourne, 63, 68
move of, to Radbourne, 40
and pairing of overlookers, 165
and reintroduction of training at Radbourne, 128
views of,
on male weavers, 133
on weaver's role, 107, 112 f.
oiler, 148 f., 153
operatives, 90
changing relationships among, 147 ff.
evolution of new frames of reference by, 125
and innovation, 212 f.
resistance of, to work study, 188
types of information relevant to, 181 ff.
views of, on innovation, 100 ff.
modification of, 102 ff.
see also under overlookers *and* weavers
overlookers,
attitude of, to lack of humidification, 45 f.
behaviour of, in new roles, 114 f., 144
changes among, 60 f.
comments of, on pairing, 167
and damaged cloth, 196 f.
effects of innovation on, 96 ff.
as 'group', 121 ff., 147, 167, 212
and innovation, 212
lack of experience of, 54
overtime by, 53, 143
pairing of, for shifts, 165 ff.
reshuffle of, 166 ff.
pooling of efficiencies among, 122 f., 214
preoccupation of, with manual tasks, 115
recruitment of, at Radbourne, 43
relation of, to weavers, 148, 152, 157 ff., 214
across shifts, 171
changes in, 97
responsibilities of, 23, 97
and running through meal breaks, 57 f.
'skill' of, 96
supervisory function of, 97
training of, 16, 122 f., 141 ff., 215
treatment of, by management, 123
views of,
on new jobs and age, 135 f.
on own role in changes, 98 ff.
and work study, 96, 122
Overlookers' Union, *see* National Association of Power Loom Overlookers
overtime, 53, 143
end of, 55

packet,
damaged, 24
definition of, 23
removal of, from looms, 93 f.
pattern of operations, 32 f.
effect of automatic looms on, 35 f.
external factors in, 37 f.
peer groups, 224 ff.
definition of, 224 f.
and industrial small groups, 225 f.
and work groups, 227
personnel manager, and changes, 44
physical barriers, and organizational scheme, 69
pick,
definition of, 23

pick—*contd.*
as measure of productive capacity, 70
'pick-at-will', inability of men to weave, 132
Pierce, 165
pirning, 20
see also under spooling department
pirns, 22
for automatic looms, 73 f.
stock-piling of, 38
preparation, drive to improve, 95
preparatory departments,
difference in, between Radbourne and Debenham, 68
extra pressure on, 58, 61, 73 f.
factors causing, 71
supervisors of, and relations with shift managers, 80 ff.
prestige, definition of, 6
production meetings, 33 ff.
daily, at Debenham, 33 ff.
at Radbourne, 34, 82, 85 f., 88
productive capacity of mill,
before and after innovation, 70 f.
method of assessment of, 70
programming supervisor, 77 f.
public transport, and shift workers, 172

quality,
at Debenham, 31 f., 38
dependence of, on factory personnel, 113
during innovation, 61 f., 87
goods, rewards for, 113
Morris's responsibility for, 31 f.
relation of, to quantity, 105
and shift working, 162 ff.
and speed, 48

racial prejudice, 120n.
Radbourne Mill, 40 ff.
administrative structure of, 18 f., 42 f., 64, 67 ff., 87
changes in personnel at, 43 f.
cloth handled at, 37
communications pattern at, 186 ff.
attitude of operatives to, 190 ff.
damaged cloth system at, 196 f.
and Debenham, 26, 32, 67 f.
desire for information at, 195
downtime at, compared with Debenham, 59, 79 ff.
history of, 12 ff., 40 f., 67, 79
introduction of automatic looms at, 14 f., 40 ff. *et passim*
introduction of shift working at, 15, 215 f.
layout of, 41, 64 f.
management group at, 82 ff.
monthly meeting at, 34
and national industry, 19
as non-automatic mill, 42
operatives from, trained at Debenham, 128
overlooker difficulties at, 54
previous changes at, 40 f.
separation of departments at, 191 f.
recruitment, begins again, 60
'research consultant', role of, 3
research workers,
involvement of, 2 f.
objectivity of, 2 f.
'resistance to change', 125, 222 ff.
and communication, 223
Robson, 165

sales department, 18
sanctions for bad packets, 112 f.
conflict of,
with constant supervision of looms, 114
with new roles, 114
and shift working, 163
satin cloth, and failure of the humidification plant, 45 f.
Saturday work, 58 f.
selection of personnel, 127 ff.
for weaving training, 129 ff.
criteria for, 130 ff.

selection of personnel—*cantd.*
by foremen, 129
by top management, 130
setts,
change in nature of, 91 ff.
size of, 23
increase in, 56
in non-automatic section, 60 f.
'sett-centred' view of changes, 98 f.
failure of new overlookers to have, 115
shaft, 21
shift managers, 153
appointment of, 49, 65, 72 f.
and managers, 82
planning of posts of, 45
relationship of, with preparatory supervisors, 80 ff.
role of, as weaving supervisors, 81 f.
and upward communication, 199 f., 203 f.
views of,
on overlookers' new position, 98
on urgency of supplies to automatic looms, 119
shift-opposites,
communication between, 170 f.
relation between, 163 ff., 168 ff.
shift work, 44, 152, 215 f.
attitude of older operatives to, 134
introduction of,
at Debenham, 16
at Radbourne, 15
and life outside mill, 172 ff.
and public transport, 172
and quality, 162 ff.
relations between operatives on, 159 ff.
and status, 161 f.
three shift, 58 f., 159 f., 176
views on,
of married men, 174
of married women, 173 f.
of single personnel, 175
for women, 54
shortage of yarn, 37 f.
shuttle, 22 f.
changing of, 92
sizing, 20 f.
department, pressure on, 53, 74, 192
junior supervisor for, 77
new machine installed, 55
supervisor of, shared with warping, 76 f.
'skill', 95 f., 109, 214 f., 220
slasher sizing, 49
smashes, 95
inability of men to repair, 132
social relationships outside work hours, 150
social structure of factory, 5 f., 150
changes in, 9 f., 43 f., 100, 219 f.
independent of culture, 200
formal and informal, 183
and 'skill', 96
social system,
changes in, 9 f.
communication allowed by, 180
definition of, 5
sort factor, 25
for nylon quality, 57
speed of looms, 91
and quality, 48
spoolers, 22, 58
spooling department, 71, 75 f.
spooling machines, automatic, 63 f.
spooling supervisor, 75
spools, *see* pirns
statistics on textile industry, 19
status,
definition of, 6
levels, increase in number of, 87
reassessments of, 210
and shift working, 161 f.
stoppages, 92 ff., 95
storage batteries,
filling of, 92 f., 94 f., 153
summary of the study, 208 f.
supervision, by weavers, 92 f.
and status, 109

supervisor-substitutes, in sizing department, 77
supervisors,
  and communications, 36, 193, 199, 203 f.
  comparison of roles and numbers of, 68
  consultations between, 36
  at daily meetings at Debenham, 33 f.
  interactions between, 84
  at mill meetings, 33
  perception of, of the innovation, 73 ff.
  and priority for automatic machines, 75
  resistance of, to work-study, 188
  views of, on daily meetings, 86

target wage, definition of, 25
Tavistock Institute of Human Relations, 3, 7
textile company,
  description and history of, 12 ff.
  production control in, 18
  programming office, 18
  and purchase of automatic looms, 14 f.
  reputation of, as 'good employer', 13
  responsibilities of, 18
textile industry, 19
  position of automation in, 206
  statistics on, 19
trainee overlookers, 142 ff.
  relationship of, with weavers, 145
trainee weavers, 138
  dissatisfaction of, with shift work, 46 f.
  join automatic personnel, 45
  relation of, with weavers, 139
  remain on fixed rates, 48, 139
  selection of, 129 f.
  views of, on training, 140 f.
  wage rise, 47
  withdrawal of, temporary, from non-automatic section, 49
  women, 131 f.
trainers, view of, on training, 137
training of personnel, 127 ff., 215, 219 f.
  selection for, 129 ff.
training programmes, 215
  for automatic section, 138 f.
  general assessment of, 145 f.
  management view of, 137 f.
  for overlookers, 141 ff., 215
    attitude of overlookers to, 143
  for weavers, 136 ff.
training schools,
  at Debenham, 16, 128, 137
  at Radbourne, 128, 136 f.
Transport and General Workers Union, 17, 116 ff.
  representative of, by-passed in negotiations, 118

unaffected groups, need to inform, 134, 223 f.
unions,
  attitude of, to the project, 1 f.
  consultation of, in work study, 24, 96
  and groups, 8, 116 ff., 151
  and redundancy of weavers, 56
  social structure of, 5
  and three-shift working, 58 f., 176
  and upward communication, 199
  and Works Council, 201
United States, rayon industry in, 37
upward communication, 198 ff.

viscose material, 'running in' with, 52

wages,
  bonus scheme, 24 f.
  fixed rate, 45, 47
warp, 20 ff.
  preparation department, effect of lengths on, 71
  preparation group, 83 f.

warp—*contd.*
production of, 76
shortages of, 61
warping, 20
department,
changes in, 75 ff.
effect on, of innovation, 74
isolation of, at Radbourne, 69
as 'skill', 96
supervisor, takes over sizing, 76 f.
warp-stop mechanism, 90
Watson, 128 f.
reverts to weaving instructor, 63, 76, 129
weaver-overlooker relations, 194, 214
across shifts, 171
change of, with innovations, 157 ff.
weaver-weaver interaction, 153 ff.
between shift-opposites, 164
opposed to automatic role, 155
and work study, 153 f.
weavers, 152
absence of, through sickness, 79
attitude of,
to absence of humidification, 46 f.
to communications, 193 ff., 198
to hooter campaign, 50 f.
criteria of behaviour of, 104 f.
development of systematic working by, 111
early arrival of, 119
effect of changes on, 91 ff., 94
and faulty cloth, 196 f.
female,
comments of, on men, 132
and shift working, 53 f., 173 f.
as group, 117 ff.
inconsistency of views on, 106 f.
increased sense of cohesiveness among, 120 f.
exceptions to, 121
and innovation, 212 f.
junior, movement of, 55
lack of information for, 194
male,
selection of, 131 ff.
shortage of, on automatic looms, 53 f.
management view of role of, 107 ff.
manual processes by, 94 f.
non-automatic, 157 ff.
relation of with other operatives, 148 ff., 159
and status, 162
overtime by, on unloading looms, 53
recruitment of, 16, 43
redundant, re-employment of, 56 f.
relationships of,
ancillary operatives, 148 f., 159
other weavers, 148 ff.
overlookers, 97, 148, 152, 171
trainees, 139
removal of packets by, 93 f.
responsibilities of, 23
and running through meal breaks, 57
sex of, views of foremen on, 133
shortage of, in non-automatic section, 60
'skills' of, 95 f.
and smashes, 95
stockpiling of pirns by, 38
and stoppages, 93
supervision by, 92 f.
training of, 16, 43, 136 ff.
union membership of, 116, 118
and use of supervisors' records at Debenham, 193
views of,
on management, 113 f.
on new jobs and age, 135
on own role, 109 ff.
on training, 137, 140
work study on, 92 ff., 153 f.
young, selection of, 133 ff.
weaving,
description of, 19 ff.
group, 83 f., 117 ff.
as 'skill', 96
weaving school, 16, 43, 53, 57
closed, 60, 129

weaving school—*contd.*
financial encouragement to join, 52
removal of trainees from, 130
reopened, 76, 129
weaving supervisors, 149
position of, in Debenham organization, 31
recruitment of, from overlookers, 84
senior, at Debenham, 31
views of,
on lack of humidification, 46
on work study, 188
weft, 20 ff.
department supervisor, effect of innovation on, 43
preparation of, 22
position of, relative to warping and weaving, 84
replenishing mechanism, 90 f.
shortages of, 61
weft and packet carrier, 148 f., 153
West (yarn store supervisor), not at management meeting, 85
wives, views of, on husbands' shift work, 173
work groups,
and culture of mill, 172, 221 f.
expansion of, in shift work, 171 f.
in Indian mill, 155 ff.
non-existence of, 151 f.
and peer groups, 227
potential members of, 151, 154
and productivity, 228
work-life, 102
work-load, 24 f., 55, 57
elevated, for new looms, 52
fixing of, in work-study, 24 ff.
for non-automatic looms, 60
work study,
and break frequency figures, 46
and communication, 187 ff.
at Debenham, 29
lack of explanation of, 189
not carried out for overlookers, 96
resistance to, 188
scheme in mill, 24 f.
and unions, 17, 24
of weavers, 92 ff., 153 f.
Works Council, 132
announcement of change to, 44
attitude of, to the project, 1 f.
constitution of, 201
and formal communication, 201 ff., 217
functions of, 201 ff.
groups in, 8, 151
minutes of, in communications system, 184 f., 186 f., 202
overlookers' boycott of, 122, 203
representative of automatic weavers on, 118 f.
social structure of, 5
and Tavistock consultants, 3
and delay in arrival of looms, 47
and unions, 201
and upward communication, 199
view of operatives of, 202 f.

yarn,
delivery of, 18
ordering of, 37 f.
preparation of, 20
shortages of, 37 f.
thickness of, effect of on spooling department, 71
Young, 43, 99, 165, 168
move of to new looms, 51
effect of, 52
view of, among overlookers, 144